FORGING A SPECIAL OPERATIONS FORCE

The US Army Rangers

Dominic J. Caraccilo

Helion & Company
Published in cooperation with the Association of the
United States Army

Helion & Company Limited
26 Willow Road
Solihull
West Midlands
B91 1UE
England
Tel. 0121 705 3393
Fax 0121 711 4075
Email: info@helion.co.uk
Website: www.helion.co.uk
Twitter: @helionbooks
Visit our blog http://blog.helion.co.uk/

Published by Helion & Company 2015, in cooperation with the Association of the
United States Army

Designed and typeset by Bookcraft Ltd, Stroud, Gloucestershire
Cover designed by Euan Carter, Leicester (www.euancarter.com)
Printed by Gutenberg Press Limited, Tarxien, Malta

ISBN 978-1-910777-36-7

British Library Cataloguing-in-Publication Data.
A catalogue record for this book is available from the British Library.

For details of other military history titles published by Helion & Company Limited
contact the above address, or visit our website: http://www.helion.co.uk.

We always welcome receiving book proposals from prospective authors.

For the Airborne Ranger

Contents

List of photographs

Preface

The site is one of the least hospitable places in all of Afghanistan. Which is saying a great deal. Unlike most bases in Afghanistan, it is devoid of locals not part of the assembled force; and intentionally so. The reason is quickly apparent when some of the occupants are seen walking across the small open area near a vehicle park. They are uniformly dressed in light grey/green skin-tight long sleeved polyester shirts and wraparound sunglasses. Protection against the cold at these heights combined with the excessively bright sunlight associated with the altitude and lack of cloud cover is critical to their wellbeing. This place is very similar to Camp 1 located on the quest to trek Everest, but with significantly more personal firepower and purpose.

Those that inhibit this base are with little doubt warriors for they look the part. Under the sleeves, the large tightly confined biceps are easily seen. The pants are a desert digital camouflage pattern finished off with scuff brown desert boots raising small dust clouds as they press on the decomposed granite and gravel that passes for dirt in this inhospitable land. Their lips are cracked and bleeding despite the heavy application of lip balm hanging in chunks around the splits. The edges of the sunglasses hide the deep fissures of the crow's feet, filled with the fine dust endemic to the area that finds a home in every possible existing crevice. Hair curls out beneath a patrol cap on one and a wool watch cap on the other. The sweat and grease on the exposed hair glistens at the edges and catches the dust in fine brown dew that collects on the tips. These are Rangers and they are serious people supporting a serious business.

To their left, under an open carport structure, are several other men. They are dramatically different. They have long beards and flowing hair and wear traditional local garb. But a closer look shows a significant difference between them and the native population. There is a group similarity to their upper body strength, relatively unlined faces and near-perfect teeth. No sign of the gross dental rot that afflicts virtually all the Afghan males-the result of a lifetime of drinking super sweet chai tea and the absence of any preventive dentistry. These are what the US Department of Defense calls Tier One forces. They are very serious people doing very serious things. Together these two groups are here to mutually conduct the most difficult and dangerous tasks that can be assigned-hunting armed humans that have a multitude of generational experiences in this land and in the fight.

The hunted and hunters frequently exchange roles depending on circumstances. For hundreds of years, the quarry has practiced its craft, adhered to Darwin's thesis and emerged as victors over the most sophisticated and technically armed societies. The latest nation state to appear has directed the tiered elements to join in the human version of the Boone and Crockett Club with the trophy game fully armed. In fact, their ability to create local leadership vacuums is crucial to the larger allied strategy. If they are not successful on a repetitive small scale, the larger engaged elements become irrelevant. Together, they and the Rangers are planning and rehearsing tonight's hunt. It's never easy. Most of the target rich environment is surrounded by naked terrain or extremely rugged access. Getting there is not half the fun.

The Rangers slowly coalesce into an informal formation. Some with weapons and some not. They gather tightly together with an assortment of watch caps, patrol caps and warm huggy covers bobbing as they converse. Spit cups are an almost universal accompaniment. Cargo pockets bulge with items essential for personal use. At a distance, it's easier to see the rise and fall of the white foam cup as they walk towards the assembly area than the body of the holders. An individual appears out of the closest door and the heads rise. As if on a signal, the Rangers move into a formation and without an order assume disciplined parade ground spacing and look attentively to the leader.

With a firm but modulated voice, the leader speaks;

"Recognizing that I volunteered as a Ranger …" When he completes his sentence, the unit, in a single voice, not loud but with firm enunciation and conviction repeats his sentence. At the final word of the first stanza, the leader looks at the center of the group and says with the same clear voice "Acknowledging that as a Ranger …" Again, the group repeats the phrase with a steady tone. In this dust-driven, ambiguous and enervating environment, these soldiers have found a lodestone to guide them and a moral compass to comfort them in the engagement ahead. The last words of the Ranger Creed softly roll across the courtyard – " … though I be the lone survivor." Almost immediately, the entire group exudes a chorus- Rangers Lead the Way! Hooah. With the Hooah on a waning declination, the group breaks up and the individuals go about their last minute preparation for the coming night's mission.

Eight hours later, the same group exits from various vehicles that have just rolled into the compound trailing a cloud of dust as they deposit their loads. Several small helicopters deposit their passengers amongst the dirt and flotsam sucked up from the compound yard. It is still dark enough to see the grey-green glow as the particles strike the tips of the rotating blades. Some, Rangers, move purposely toward the hooch they assembled at the previous dusk. Others, in native dress, many with thick beards, ambled toward their portion of the compound.

The Rangers, some quicker than others, gather in an informal assembly. Some move slowly, more shambling than erect, bent over with either gear or exhaustion or both. They are now fully equipped with all the killing tools of their combat equipment, night vision devices, commo gear with embracing wires and antennas, Kevlar protection and with some butt packs now loosely closed absent their original contents. Their load bearing straps are arrayed with a variety of ammo pouches, lights, grenades and the miscellaneous comfort items soldiers habitually carry. Their heads, now sweat and dust streaked, are either in Kevlar or watch caps and the movements display the exhaustion of the night's activities. Rivulets of grimy water course down exposed necks making small streams of exposed flesh. The look of freshly born combat on their faces is confirmed by the presence of several Rangers who have bandages on arms, legs or necks. It's been a long night. Weapon muzzles are coated with a light cast of dust, the twilight still too dark to render meaningful color. An occasional passing light beam activates the Glint tape of a Ranger for a moment before it passes.

As if by osmosis, the group coalesces dissolving into a reasonable facsimile of a formation once again. The leader stands in front and begins what has become a daily ritual of recovery from the evening's program "… Recognizing that I volunteered as a Ranger …" On the initial words, the group automatically assumes a tighter formation, straightens up their heads and alignment and repeats the stanza … "Recognizing that I volunteered as a Ranger … " The voices are somewhat more muted than earlier but occasionally a Ranger will become particularly loud or concise in repeating a specific set of words as if they were cathartic and an antidote to what he had just experienced.

Finally, the last stanza of the Creed is spoken " … though I be the lone survivor!"

The group raises its voice several degrees, spits the words out with a single breath and without orders, and breaks away to their various home stations for cleanup, recovery and rest. The Rangers are home for another twelve hours. The sounds of the Creed dissipate in the cold, dry wind, but are not lost to either those that spoke them or to those that faced the full measure of their meaning. The Ranger Creed is more than words, they are a lifestyle. And they materially assist those charged with taking it from others.

Keith Nightingale
U.S. Army Ranger
February 2, 2015

Introduction

"Once more into the breach … dear friend"
William Shakespeare, *Henry V*

FORGING A FORCE

Forging a special operations force or any force for that matter is no small task. To "forge" is to build, to form, to create and in many cases to draw on desirable qualities, attributes and characterizations of something that already exists in order to fashion how an organization is formed. To develop an organization or anything significant requires a need to change or, at the very least, a great desire to design and build something new. But it can't be done in a vacuum. A tried and true precedence helps in developing a decision to instigate something new. And this effort is not without its own precedence. Armies and organizations throughout history come and go. Some have a lasting impact on society; others fade away without notice.

The author has been fortunate to experience these types of changes first hand both in his military career and in the years following retirement. Having stood up a new airborne battalion a year prior to a combat jump into Iraq, leading the transformation of the 101st Airborne Division in the year proceeding a yearlong deployment to Iraq, and then leading the training and readiness transformation of the United Arab Emirates' Land Forces in the year following retirement from the US Army has given the author a firsthand look at the challenges associated with this type of effort.[1]

Forging the US Army Rangers as a special operations force in the 1970s had its own set of challenges. Fortunately for our Army, those challenges were surmounted and today the 75th Ranger Regiment is the world's premier small unit infantry and special operations force capable mastering a full array of infantry tactics and special operations mission sets.

The Army of 1974 was struggling. The Great Green Wave, a synonym many of the conscripts called the US Army, was in terrible shape as it returned from Vietnam. The officer corps who saw their men's lives wasted in Southeast Asia by misguided civilian strategists in Washington—began to say, "Never again." The unpopular Vietnam conflict was waning and many believed that a conscript Army, full of indiscipline had an open resentment to the nation's leaders for how they prosecuted the war, was a leading reason that the US had failed in the overall campaign.[2] Absences without leave (AWOL), disrespect, drug problems and poor leadership at the junior leader level were just some of the issues in the Army at the time.

After years of dealing with much consternation, the US administration discontinued the draft in 1973, moving to an all-volunteer military force and for the Army it was termed the all-volunteer Army (VOLAR). This decision ended the mandatory enlistment and the eventual issues brought on by the draft began to fade away.

To counter the prevalent disciplinary and readiness challenges brought on by the mandatory draft, military leaders in 1974 insisted that the Army stop accepting social misfits even after the end of the draft in order to fulfill recruiting quotas. Additionally, the senior leaders were determined to rid the Army of racism and took significant educational and disciplinary measures necessary to do so.

Leaders worked vigorously to stamp out drug abuse, both by rehabilitation programs and by discharging those who couldn't be helped. Senior military leaders convinced the authorities, military and civilian, to make life in the army something more than an ongoing, low-grade misery in the barracks. They stressed educational opportunity for recruits, and they rebuilt the backbone of the Army, the career noncommissioned officers (NCO) corps, whose ranks had been particularly decimated by Vietnam. They instituted graduate schooling programs to increase the educational level of their officers and implemented an advanced command school program to enhance their professional military education (PME).

Above all, military leaders during this time, were self-critical and learned to be even more thoughtfully self-critical in the revamped programs. They took risks in challenging superiors and they challenged those peers who couldn't cut it. The best leaders found in the ranks during this time were prepared to resign rather than see out their careers in comfort. This was true even if the latter meant risking a repeat of the disaster through which they had lived as young officers. Mostly, these leaders were determined to train realistically and to increase soldier performance standards and, as a result, maximize the discipline of the US Army Soldier

Revamping the Army was no small task. For years a conscript force, while performing selflessly and admirably, had been a burden on the Army for the Soldiers became less and less motivated to perform; especially in an unpopular war where troops counted the days until they could "get out". This type of attitude led to mediocre performance and ill-disciplined formations while the country was at war in Vietnam. A malcontent Army composed of conscripts led to a lack of respect for rank and authority and, in turn, an inability to lead small unit organizations in some measured fashion. In turn, the experience to train to a high standard of performance was lacking throughout the Army. Clearly there had to be a change.

Today the volunteer force arguably makes up the best Army that has ever existed in the history of mankind. What makes it unique and special is that every member of the US Army chooses to be a part of it. No one forces our soldiers to serve in the Army today; they all volunteer. Most impressively, Rangers are volunteers three times over– they volunteer to enlist; to jump from airplanes; and finally to become a Ranger.

The desire to go from a conscription force to an all-volunteer force (AVF) had been an ongoing debate with the political leadership throughout the 1960s and 1970s. The decision to eventually implement this change was a decisive move.[3] But, it wasn't without challenge and much effort.

During the 1968 presidential election, Richard Nixon campaigned on a promise to end the draft.[4] He first became interested in the idea of an all-volunteer army before becoming president. Reflecting back on decades of experience with the AVF suggests four broad reasons for its success.

The first reason was attention and leadership from top management. The AVF would not have been established at this time without the leadership of President Nixon. Within weeks of taking office in 1969, he began the planning process and announced the formation of the Gates Commission n to advise him on establishing the all-volunteer force.[5]

The Gates Commission was headed by Thomas S. Gates, Jr., a former Secretary of Defense in the Eisenhower administration and, at the time, the Executive Committee Chairman of the Morgan Guaranty Trust Company. Its membership included other distinguished businessmen; former military leaders Maj. Gen. Alfred Gruenther and Lt. Gen. Lauris Norstad; and a university president W. Allen Wallis. Distinguished and influential scholars included economist Milton Friedman. Roy Wilkins, Executive Director of the National Association for the Advancement of Colored People (NAACP), was a member, as was Georgetown University law student Steven Herbits. The diversified makeup of the commission clearly was intended to generate assurance that the decision to end the draft was reasonably based.

Gates initially opposed the all-volunteer army idea, but quickly changed his mind during the time he spent on the commission. The Gates Commission issued its report in February 1970, describing how adequate military strength could be maintained without having conscription. The existing draft law was expiring at the end of June 1971, but the Department of Defense and Nixon administration decided the draft needed to continue for a brief period longer. In February 1971, the administration requested of Congress a two-year extension of the draft, to June 1973.[6]

A second contributing factor for the success of the all-volunteer Army was the use of quantitative analysis to test, adjust, and evaluate AVF policies in an effort to optimize the result. Well-designed, policy-relevant research was conducted to measure job performance for each skill set and pay grade and a determination of the optimum mix of quality and cost. This analysis resulted in the proficient and committed AVF that serves the nation today.

The third factor for the success of the all-volunteer Army was the availability of adequate financial resources. The defense budget had to be large enough to support pay raises to keep pace with both inflation and civilian-sector pay increases; to provide resources for advertising, recruiters, bonuses, and educational benefits; and to fund the military retirement program and quality-of-life initiatives.

The final factor was the need to develop programs for attracting the necessary type and number of recruits. The AVF's focus on quality is an obvious choice of priority. To attract high-quality recruits, the services had to develop appropriate marketing strategies and advertising programs that explained the benefits and opportunities of military service. The military learned that it had to offer money for education, bonuses to enlist and maintain certain occupations, and enlistment tours of different lengths. It needed to develop career opportunities that had civilian relevance and were a good preparation for adulthood. All services to include the Army also had to develop a professional, highly trained, and motivated recruiting staff.

The Army learned that reenlisting the most capable members was the key to creating a truly outstanding force. This is something the 1st Ranger Battalion took to heart in their initial effort to organize during the standup. Besides good pay, careerists demanded quality-of-life benefits such as adequate housing, child care, health benefits, family advocacy programs, and military stores. It was crucial that the services become "family friendly." Most importantly, military leadership learned over the years that new soldiers wanted to be a part of something bigger then themselves in the form of high quality organization. As a result, the implementation of performance based standards into the services in all that they did became the universal reason the all-volunteer force eventually succeeded. And it began with the establishment of the 1st Ranger Battalion in 1974.

These views were reinforced by the findings of the Gates Commission. The commission addressed key military-manpower issues, including supply and demand, attrition and retention, and the mix of career and non-career members in the context of management efficiency and personal equity. It concluded that the nation's interests would be better served by an all-volunteer force than by a combination of volunteers and conscripts. Because there was opposition to the all-volunteer notion from both the Department of Defense and Congress, Nixon took no immediate action towards ending the draft earlier in his presidency.[7] But by 1971, President Nixon signed a new law to end the draft and put the selective service structure on standby. After a two-year extension of induction authority, the end of the draft was formally announced in January 1973. Nixon also viewed ending the draft as an effective way to undermine the anti-Vietnam war movement since he believed the nation's affluent youth would stop protesting the war once their own probability of having to fight in it was gone.[8]

Meanwhile, military pay was increased as an incentive to attract volunteers, and television advertising for the US Army began. With the ending of active US ground participation in Vietnam, December 1972 saw the last men conscripted, who were born in 1952 and who reported for duty in June 1973. On February 2, 1972, a drawing was held to determine draft priority numbers for men born in 1953, but in early 1973 it was announced that no further draft orders would be issued. In March 1973, 1974, and 1975 the Selective Service assigned draft priority numbers for all men born in 1954, 1955, and 1956, in case the draft was extended, but it never was.[9]

LEADING THE WAY

It is clear that the mid-seventies, post-Vietnam era was replete with change for the US Army. As the war in Vietnam ended it became clear that the force that struggled to win the war had to transform or become as impotent as it was during the War. Critical to the transformation of the US Army into a professional volunteer force was the implementation of a high performance, standards-based outfit. This outfit had to be the standard bearers for the rest of the Army. The force of choice was recognized as the 1st Ranger Battalion which would eventually evolve into an entire Ranger Regiment as the decade after Vietnam unfolded.

The introduction of a permanent Ranger element that instilled the high standards in both discipline and training was the result of the quest for a standard bearer amongst the ranks. In fact, one of the reasons for the formation of the modern day Ranger Battalion was to provide a change agent for the rest of the Army.

This book provides insight into the implementation of the Ranger mentality and how it was infused into the US Army. It shows the mindset of this elite soldier as seen through the eyes of the founding fathers; the plankholders of the modern day Rangers.

Dating back to Rogers Rangers, specialized commando units have been formed in most of our country's conflicts. These temporary units were targeted at specific missions related to that conflict, then disbanded after the end of hostilities.

In 1974, with the Army coming out of the throes of Vietnam, with its reputation and morale at low ebb, General Creighton Abrams, Chief of Staff of the Army (CSA) and former Commander, US Military Assistance Command in Vietnam (COMUSMACV), directed the formation of a Ranger Battalion (1st Battalion, 75th Rangers or 1/75th). His guidance became known as the *Abrams Charter*.

Chief of the
Staff of the
Army General
Creighton
Abrams.

General Creighton W. Abrams was a 1936 graduate of the United States Military Academy (USMA) at West Point. An armor officer, he served as the commander of the 37th Tank Battalion in General George S. Patton's Third Army in World War II. He gained fame when he led the battalion in relief of Bastogne during the Battle of the Bulge. General Patton is quoted as saying: "I'm supposed to be the best tank commander in the Army but I have one peer – Abe Abrams. He is the world champion."[10]

Abrams also commanded the 2nd Armored Cavalry Regiment in the post-war occupation of Germany and the 3rd Armor Division during the Berlin Crisis.[11] In Korea, General Abrams was the Chief of Staff of three different Corps.[12] He also served as the Vice Chief of Staff of the Army and eventually succeeded General William C. Westmoreland as the commander of all United States forces in Vietnam. Two years into his tenure as Chief of Staff of the Army, Abrams fell ill with lung cancer, an ailment that eventually took his life. He is the only Chief of Staff to have died while in office.[13]

When all the fog lifted the *Abrams Charter* was established with the formation of the 1st Ranger Battalion. Abrams directed that this battalion be the finest infantry unit in the world; that it must set the standard for the Army; that veterans of this battalion would eventually serve with regular Army units and spread those standards across the Army. The effort to infuse this type of mentality across a force by spreading the talent developed in specialized units had been done in the past and I will discuss this more in subsequent chapters on the history of the Ranger.

While this quest sounds like a simple concept and a decision easily made, there is more to this story and this work will reveal the specifics. Nonetheless, the *Abrams Charter* came to fruition in a way that was never expected.

While earlier Ranger-type units were disbanded, the original 1st Ranger Battalion evolved into a Ranger Regiment of three battalions that is now considered the world's premiere strike force. Since the formation of the first permanent Ranger force in 1974 its units and, perhaps more importantly, the individual Rangers (those that have had experience as both Rangers and those that are Ranger qualified) have been in every combat operation the nation has fought since Vietnam to include Grenada, Panama, Desert Strom, Kosovo, Iraq (Operation Iraqi Freedom) and Afghanistan (Operation Enduring Freedom).

Forty-plus years after the original formation of that first battalion, the modern-day Rangers are an integral part of our Army's capabilities. This work describes the genesis of the establishment of the unit, its standards and performance measures, and most importantly the men who were responsible for the formation of this force.

However this book is more than just a history of how the initial Ranger force was forged. It is about the psyche of a soldier we call Ranger. It is about those that forged such a fighting machine. It is about the selfless sacrifices made by courageous and creative men so that an Army, on the brink of failure in a post war era, could find itself once again and be forged into a formidable force. It is about how the Ranger psyche has had an impact on our soldiers and their Army for the past 40 years and how that impact led to the successes seen today.

As stated earlier, the mid-70s saw sweeping changes throughout the military that affected the entire force for years to come. The premise of this work is that one of the most important changes; if not the most important of them all during this time was the establishment of the US Army Rangers Battalions within its ranks.

Ranger units have been active since before the Revolutionary War starting with the acknowledgement of the very first Ranger force in the 17th Century. Progressively involved in 18th and 19th Century warfare thru the monumental successes of the World Wars, Korea, and Vietnam and beyond, Rangers and Ranger formations have been involved in every major US conflict leading to and including protracted ground war in Vietnam. However, in every single incident the Ranger units were always disbanded or transformed after hostilities ceased.

The post-Vietnam era saw for the first time the inclusion of the Ranger Force as a permanent entity in the US Army. As a historical examination of the US Army Rangers in this era this work examines the difficult decisions leaders made in the establishment of a permanent Ranger organization; their philosophy in how to train and lead; the adherence to a discipline evoking Creed; and a "big Army" Charter that paid big dividends to the entire Armed Forces in the years to follow

To support this amazing story this work explores the Rangers' role in key US engagements, their evolution on the battlefield, and how they continue to serve in today's wars in the Middle East. Today's Rangers are, at once, an elite infantry unit and the epitome of a special operation formation. This volume takes a look back at how this elite fighting force was formed and its story told through the lens of those who forged such a force.

This work is replete with dozens of interviews from those that created the initial Ranger Force in the post-Vietnam Era which led to the elite fighting force that it is today. Unique in this work is the exploration of the mindset of individuals involved in developing such a force when General Creighton Abrams conjured up the thought to develop America's premier fighting unit. The first Army Ranger Battalion was a necessity for the US Army's survival after Vietnam. Its formation and institutionalization established high standards; a meticulous approach to training for war fighting; and a leadership workshop for the rest of the Army to emulate.

WHY ANOTHER RANGER BOOK?

While it is important that the reader understands the entire history of the Ranger, this book is more than just a history of how the initial Ranger force was formed. It is about those attributes that form the Soldier we call Ranger. It is about those leaders that forged such a fighting machine. It is about the selfless sacrifices made by courageous and creative men so that an Army, on the brink of failure, could develop itself once again as a top fighting organization. It is about how the Ranger spirit has had an impact on our soldiers and their Army for the past four decades and how that impact led to the successes we see today.

The Ranger Organization imbedded in the US Army in the post-Vietnam era was a critical addition at the time in order for the Army to survive as a professional fighting force. The infusion of this select organization into the "big Army" in the 1970s and 1980s has proven its worthiness for today's US Army represents the best fighting force in the world.

With the United States recently engaged in several major conflicts in the world, this book is not just for those that serve in Ranger units or military history buffs with a niche for Ranger history. It is written with the hope that it will would bode well with lay readers interested in the direction their nation is taking and what type of combat forces it utilizes—especially in those areas where special operations forces are operating which is most areas of combat.

While there are many books on Rangers already in print this book is different. The reader can see in the exhaustive bibliography at the back of this work to see the magnitude of tomes on the market or in the archives. Of note, John Lock's *To Fight with Intrepidity: The Complete History of the US Army Rangers 1622 to Present* and Ross Hall's *The Ranger Book: A History 1634-2006, "Not for the Weak or Fainthearted"* were integral to the background matter for this work. Lock's work is a replete history of the Rangers leading up to the end of the 21st Century. Mir Bahmanyar's *Shadow Warriors: A History of the US Army Rangers* is another detailed account of the Ranger history covering the time period up to an including its publication date of 2005. In it, Bahmanyar provides a general coverage of the 1st Battalion activation in 1974.

It is the author's intent that this book accompanies and adds value to these and other works. So much so that a definitive Ranger history can be told with a focus on firsthand accounts never before available to authors in describing the psyche of the Ranger during the post-Vietnam era through the eyes of the plankholders who found the modern day Ranger Force.

With war on the mind of the media, public interest in America's military power remains high. Furthermore, the interest in the elite forces resident in our US Military has always been heightened at the time of war. Lay readers, academicians, political leaders and policymakers, strategists, students and members of our military will undoubtedly find great interest in this work for it defines the foundation of our American Fighting elite.

This book also provides a snapshot in time of what it was like to be soldier in 1974 at the end the conscript Vietnam era Army. The author attempts to weave colorful depictions of the Army at that time which led to the effort to reorganize and reform America's elite fighting forces that ultimately led to a universal increase of standards and discipline throughout the Army.

This work includes a supportive history of the Rangers dating back to the 17th Century providing the reader a firm understanding of the evolution of this great force. Robert Rogers who led the famous Rogers Rangers; Francis Marion otherwise known as the "Swamp Fox;" John Mosby of American Civil War fame; Darby's Rangers of World War II; the Ranger Companies in Korea and Vietnam; and the modern day era Rangers all come to light in this work. However, the focus and the heart of the book is on the establishment of the 1st Ranger Battalion in 1974 and the men who forged the force at that time.

The firsthand accounts and interviews with the founding fathers of the modern day Ranger, most notably Major General (retired) Kenneth C. (KC) Leuer and his subordinate commanders and other Rangers at the time provide the basis for the story that evolved in the early 70s. Beside Leuer's input, Brigadier General (retired) Ron Rokosz, Colonel (retired) Keith Nightingale and Major (retired) Todd Currie provided hours and hours of research from personal interviews to make this work come to fruition. They deserve the thanks of all Rangers past and present. These interviews by the original members and those that helped form the General Abrams' vision are woven into the author's depiction of how the initial Battalion was formed.

Compelling vignettes from the early days in 1974 and those taken from current conflicts in Iraq and Afghanistan provide the reader an ability to understand what it is like to be a modern day Ranger and how they became as good as they are today. These vignettes also help the reader understand the Ranger psyche and its impact on the rest of the Army over time. Finally, the author takes a brief look into the future of the Ranger Regiment and those supporting special mission units since the Rangers permanent activation in 1974.

In the end, this is not just another chronological regurgitation of the Rangers and their past. It is an exploration of the mindset of the way the Rangers were reborn in the early 1970s. It brings to the forefront the courageous and persistent exploits of a special breed of men for whom the phrase "Rangers Lead the Way" truly means something more than words. It's a way of life.

1

Ranger History through the post-Vietnam War Era

Then I heard a voice of the Lord saying, "Whom shall I send? And who will go for us?"
And I said, "Here I am, Send me!"

Isaiah 6:8

This chapter provides an ample history of the Rangers and Ranger units since the seventeenth century. It was in this era that the modern-era Rangers claim the founding of their lineage up to and including the era immediately following the US conflict in Vietnam. It is important for this work to look at the past to get a better understanding of how the Ranger Force of the present is organized and operates. Much of what happened in previous centuries has had a large impact on the formulation of what it is to be in a Ranger organization today.

The genesis and chronological history of the US Army Ranger is arguably in question, but there is little doubt that it predates the US Revolutionary War. The labeling of the term "'ranger'" is used in thirteenth-century England to describe a far-ranging forester or borderer.[1] By the 1600s the term Ranger began to serve as a label for irregular type military organizations.

The actual word comes from the Frankish "hiring," meaning circle, or ring, which describes the patrolling and flanking movements, which are often curved so that the attacking force can get in behind the enemy. English foresters were "keepers of the royal forest," routing poachers by using a widening circular movement to locate intruders, *ranging* outward in widening circles.[2] Ranger presence in the Americas mostly likely began when English colonists established defenses to fend off Indian attacks on Virginia settlers on March 22, 1622. It was then that armed men roamed or "ranged" perimeters to provide early warning resulting in the term "ranger."[3]

US Army Ranger lineage falls into two distinct periods. The early years encompass the pre–American Revolution with the likes of Robert Rogers, the American Revolution with Francis Marion, and the American Civil War with John S. Mosby. The second era is commonly referred to the "modern" ranger era, which began with the formation of Darby's Rangers in World War II and led to the post–Vietnam War era. This chapter provides a historical overview of both of these eras.

THE EARLY RANGERS

The deep roots of Ranger heritage were established in Colonial times more than a century before the Revolutionary War to counter expanding threats against the colonies by Native Americans. In letters written in 1609–1610, colonial military leader Capt. John Smith described Indian tactics; "They never fight in open fields but always among reeds or behind trees taking their oportunitiy to shoot at their enemies till they can knock another arrow they make the trees the defense [sic]."[4] These natives avoided pitched battles and open field combat for a more clandestine method of raiding as a tactic. Rangers learned early on that to emulate this type of fighting which proved successful. The Indian became one with his environment. Using hunter's skills, he took advantage of cover and conceal-ment to close with the enemy and killing the enemy while taking as few casualties as possible.[5] As a result, colonists thought it wise to put aside their less useful European tactics to emulate the ways of the Indian. From this attitude and effort was born the American Ranger.

As Robert Black writes in *Ranger Dawn: The American Ranger from the Colonial Era to the Mexican War:* "Absorbing his environment, the Indian became one with it. He used the hunter's skills, took advantage of cover and concealment to close with his enemy and made the most of his primitive armament of bow and arrow, club, spear and knife. Success meant killing the enemy while taking as few casalties as possible."[6] As a result, colonists thought it wise to put aside their less useful European tactics to emulate the ways of the Indian.

The history of the American Ranger is long and colorful. Over time, the Rangers' tactics, techniques, and procedures have kept pace with emerging technology. The earliest mention of actual Ranger operations comes from when Captain Smith who wrote in 1622, "To range this country of New England, I had but eight [men], as is said, and amongst their brute conditions I met many of their encounters, and without any hurt, God be thanked."[7]

The ideas behind ranging date well before the seventeenth century colonists' encounters with the Indians. During the Middle Ages, *Rangiatorem*, or Rangers served the English King in the 14th Century.[8] However, the first Ranger acknowledged by name by most historians is Edward Backler. In 1634, he was hired to serve as a Ranger in the upper Chesapeake Bay area in present-day Maryland. Here politician William Claiborne hired Backler to patrol and give warning in the Virginia territory to provide information on the movements of his enemies.[9]

The use of the Rangers in the southern colonies became commonplace in the 17th and 18th Centuries. In the north the colonies began using Rangers for defense against Indians when in 1670, the colony of Plymouth (part of modern Massachusetts) maintained a unit under Thomas Willet. During King 'Philips' War (1675–1678), also known as the Metacomet War, Plymouth and Massachusetts raised and maintained Rangers for both defensive and offen-sive purposes.[10]

Around 1700, there were at least two companies of Rangers patrolling between what are now Washington and Baltimore. It is important to note that though these men were under British authority, they were technically Continental Militia and did not perform operations for the British as regular troops.[11] These independent ranging companies organized into settlers and friendly Indians were led by colorful leaders, such as Benjamin Church.

THE 1600S AND BENJAMIN CHURCH

One of the first recognized Ranger leaders was a loyalist named Benjamin Church. Church commanded one of the first Ranger companies formed in 1675 at the Plymouth Colony and organized it to defend the colonists from hostile Indians.

Church was an effective leader who learned to combine Indian tactics with European discipline and weaponry. Some 80 years before Robert Rogers of Ranger fame wrote his ranging rules, Church was writing on the importance of not traveling the same route twice. He also internalized the way the Indians moved with their men dispersed out rather than in a body as the English did. These movement techniques and other Indians tactics such as attacking at night, separating captives for interrogation, and stealthy reconnaissance methods are the basis of how the Rangers of today operate.[12] The Rangers of the colonial era set in motion the historical precedence for the selfless discipline infused in the contemporary force. Like the Rangers of today, these Rangers were not only attracted to the dangers and hardships of this type of service they were all volunteers.

Unlike other militia organizations of the time, Church incorporated friendly Indians into his ranks, where they were treated like fellow soldiers. Some of the Indians even converted to Christianity in settlements before the war. These were known as Praying Indians.[13]

Church was known for his many expeditions into Acadia, which included present day Maine, to hunt down King Philip in 1676. Severely obese and progressively treating Indian prisoners in an inhumane manner, Church continued his raids but the authorities lost confidence in his leadership. It got so bad that in 1692 orders were given to secure the safety of any Indian captive taken by Church and his men.

Executing his final expedition into Acadia in 1704 Church took prisoners and destroyed the infrastructure leaving only five houses standing in in the area. One of the prominent prisoners taken in this raid was the Acadian leader, Noel Doiron.[14] This final expedition brought Church, one of the founding fathers of the modern-day Rangers, career to an end. One of his many legacies was his memoir *Entertaining Passages Relating to Philip's War,* published in 1716 and is considered the first American military manual.

THE 1700S AND ROBERT ROGERS

By the 18th Century, Ranger type units became very good at raiding, ambushing and conducting reconnaissance on hostile Indians and other foe. Most of the remaining northern colonies began to use Rangers to protect their colonies and surrounding frontiers making the use of Rangers commonplace. Wide open in 1700, Georgia was a key contested area, lying in between the two colonies. In South Carolina, the ranging system was set up with designated companies operating at and in between static forts along the various rivers terminating on the coast.[15]

The next noteworthy example of an evolving Ranger concept arose in the conflict known as the French and Indian War being fought on the American continent and ran parallel to the Seven Years' War (1755–1763) that took place in Europe.

European warfare was the antithesis of the new style of warfare witnessed and developed in the America. Standing shoulder to shoulder firing volleys into each other's ranks became an antiquated way of soldiering for it wasted lives. Frontiersmen in the North America recognized it as such. However, the idea of firing from cover and concealment was implausible to those who soldiered during this era. The challenge of dealing with the guerrilla warfare tactics of the North American Indians, however, required a change and a leader innovative and brave enough to make that change. Robert Rogers was that man.

Rogers was a rugged frontiersman who was immortalized in American literature as the main character of Kenneth Roberts's classic novel *Northwest Passage*.[16] Founded on March 23, 1756, Rogers' Rangers were paid by the British Army, but were subject to military discipline imposed by the colonists. Their mission was to harass the enemy wherever they could.

Eventually, Rogers assembled 10 companies to fight against the French and their Indian allies. When they went on patrols, the Rangers moved in small-sized elements, and when contact with the enemy was made, they would disperse, conduct their missions, and rally after at different locations. These procedures were so effective that Rogers decided to put the Ranger training and standard operating procedures (SOP) onto paper. The Standing Orders and ageless rules are to this date taught to new Rangers of all ranks as they are challenged in the Ranger Indoctrination Program (RIP) and the Ranger Orientation Program (ROP).[17]

Rogers was ideally suited to lead these warriors. Although there was an insistence on discipline and rules, Rogers tended to be an unorthodox leader leading unorthodox men fighting an unorthodox style.[18] To emulate the successes achieved by American Ranger elements, the French and Indians formed special anti-Ranger companies. This effort was very similar to the North Vietnamese development of a reaction force to fend off the American Ranger Long Range Reconnaissance Patrol (LRRP)/Long Range Patrol (LRP) teams seen later in Southwest Asia.

Operating as a Ranger in the winter challenges in the Northeast required hardened men who could survive the elements. Regardless of the threat—terrain, weather, or enemy—Rogers' Rangers was a formidable force striking at the enemy in their own areas of operation in the form of deep raids which became a Ranger trademark. This reputation is best in the following rules Rogers penned and live today as a set of rules for all Rangers seen in Appendix B – Rules of Discipline and Standing Orders.

On September 14, 1757, Rogers's "Ranging School" was officially authorized. Its first group of students were British Cadet volunteers. To structure the training, Rogers drafted the 28 tactical rules shown earlier as "Rogers' Rules of Discipline." The origin of the Standing Orders remains in question. There is some question when they were actually written and some believe they are derived from Kenneth Roberts's 1936 *Northwest Passage* novel. Regardless, his standing orders stand today – see Appendix B – Rules of Discipline and Standing Orders.

THE REVOLUTIONARY WAR THROUGH THE NINETEENTH CENTURY

Veterans of Rogers' Rangers played a major role in the Continental Army during the American Revolution (1775–1784) which saw the wide use of Rangers both individually by state and at the Continental Army level. On June 14, 1775, with war on the horizon, the Continental Congress ordered six companies of expert riflemen be immediately raised in Pennsylvania, two in Maryland, and two in Virginia.[19]

THOMAS KNOWLTON

One of the most famous Revolutionary War Rangers was Thomas Knowlton. Knowlton is commonly considered the first military intelligence professional. His elite, personally selected group was America's first official spies and known as "Knowlton's Rangers." This force of less than 150 handpicked men was used primarily for reconnaissance to include Capt. Nathan Hale who is known as one of his most famous American spies. Today's "1776" on the modern US Army's Intelligence Service seal, refers to the date of Knowlton's Rangers formation.[20] In 1995, the Military Intelligence Corps Association established the Knowlton Award, which is presented to individuals who have contributed significantly to the promotion of army Intelligence.[21]

MORGAN'S "RANGERS"

In 1777 Washington organized a group of expert riflemen into what was known as The Corps of Rangers led by Col. Daniel Morgan. This unit was called "Morgan's Riflemen." Their service ran from 1775 to 1781. Some of their most famous battles were fought at Freeman's Farm during September 1777 and at the Battle of Cow Pen during January 1781 against General Cornwallis's best British troops. According to General Burgoyne, a famous British general, Morgan's men were "the most famous corps of the Continental Army. All of them crack shots."[22]

THE SWAMP FOX

Another hero of the Indian Wars who went on to fight for the American's Revolution was five-foot-tall Francis Marion, otherwise known as the "Swamp Fox." Most of the Revolutionary War heroes were not the saints that biographers portrayed; and Francis Marion, was no different. Marion was clearly not cut from eighteenth-century nobility. As a soldier his ability to recognize early on how the Cherokee used the landscape to conceal themselves before conducting lethal ambushes, resulted in his application of these tactics decades later against the British.[23]

Partisans led by Marion numbered anywhere from a handful to several hundred. They worked with and independently of Gen. George Washington's Army. Many a Brit attempted to pursue the Swamp Fox, including British Colonel Tarleton, who pursued him for 25 miles one time only to arrive at a point that was impassable. Tarleton halted and cursed something like "the damned fox, the devil himself could not catch him." From this encounter Marion was thereafter known as the Swamp Fox.[24]

Operating from the Carolina swamps, the Swamp Fox and his men disrupted British communications and operations. The Swamp Fox served his cause from 1775 to 1781, and like most Rangers in the years to come, he used many of Rogers's standing orders during engagements in and out of the swamps of Carolina.

The 2000 movie *The Patriot* depicted a fictional character named Benjamin Martin who was introduced in the story as a kindhearted warrior and family man. While this was not an accurate picture of Marion, and his exploits were exaggerated in true Hollywood form, the Swamp Fox legend was presented for a whole new generation to emulate and appreciate.[25]

POST-REVOLUTIONARY RANGERS

At the end of the Revolutionary War, all existing Ranger units were disbanded, a tradition observed after each major war for the next 170 years. But once again, the War of 1812 found units and soldiers operating as Rangers during the three-year conflict with the British. Throughout the War of 1812, Rangers patrolled the frontier from Ohio to western Illinois on horseback and by boat.

The famous eighteenth-century pioneer and Ranger Daniel Boone fathered two sons, Daniel Morgan and Nathan, whom commanded Ranger companies during the War of 1812. Notably, Nathan commanded a company of horse Rangers that became part of the Missouri militia.[26]

In the post 1812 era, the United States' focus was shifting to expanding to the west, where the need for Rangers for protection with their newly formed frontier establishments was paramount. By 1832, the United States maintained a 600-man element of mounted Rangers in the new frontier. Although the western commander, Gen. Zachary Taylor, had a poor opinion of Rangers units, he found them to be vital as scouts during the Mexican War.[27] Throughout the nineteenth century, Rangers were the nation's volunteers and had the reputation for being elite and capable of employing a set of special skills unfounded with conventional army soldiers.

Without formal training and poor pay, the early Rangers mostly comprised of those who were comfortable living in the elements. Independent and self-armed carefree men with leadership skills and the ability to operate on their own with little support were the most successful then and now.

Notably was Illinois native Abraham Lincoln who in 1832 was a member of the State Frontier Guard, whose members were called Rangers. Although he never saw combat, Lincoln served in Elijah Illes's company during the period May 29–June 16, 1831. Lincoln then reenlisted as a Ranger in Jacob Early's company, known as an "independent spy company," where he served from June 20 to July 10, 1831.[28]

Prior to his election as the 16th president of the United States, Lincoln referred to his time in service fondly, noting his selection as captain as one of the proudest moments in his life and his service has been referred to as a "shaping circumstance in his life."[29]

THE CIVIL WAR RANGERS

The American Civil War (1861–1865) featured Ranger units serving in both the Union and Confederate Armies. These "partisan" Rangers, like a unit called Quantrill's Raiders commanded by William Clarke Quantrill, were authorized by their respective armies but raised by the states.[30] In the end, the Civil War, mostly in the Confederacy, saw a total of 428 units designated as Rangers during the conflict with the majority being Ranger-style units in name only.[31]

MOSBY'S RANGERS

In 1862 the Confederacy authorized six regiments, nine battalions, and a number of independent companies to serve as Ranger units in eight different states. Many of these units were ill disciplined and untrained.[32] The exception to this was perhaps the Rangers under the command of John Singleton Mosby, affectionately known as the "Gray Ghost" for his ability to move his unit, known as "Mosby's Rangers," clandestinely around the battlefield.

Mosby believed that by resorting to aggressive actions he could compel his enemies to guard a hundred points.[33] He was greatly influenced by Francis Marion and used many of his and other Ranger tactics, lessons and experiences to develop an aggressive fighting force emulated by Rangers today. He had the mindset that if you are going to fight, then be the attacker.

The concept of a Ranger-style cavalry unit was first voiced by Mosby in December 1862 serving with Jeb Stuart's cavalry in northern Virginia. Mosby approached Jeb Stuart with the idea of not bivouacking during the winter months like most armies of the time. Instead, he suggested continuing operations and conducting guerrilla activities in the Loudoun County during the months of "hibernation."[34] This was a novel idea at the time and changed the way soldiers fought 24/7 for years to come.

After several battles in the closing years of the war and two serious wounds, John Singleton Mosby disbanded his Ranger unit on April 21, 1865 in Salem, Virginia. Upon Lee's Surrender at Appomattox, epitomizing the first stanza of the current Ranger Creed, "Surrender is not a Ranger word," he refused to formally surrender.

JOHN HUNT MORGAN

The first time an organized cavalry squadron was formed was during the Civil War, and its leader was a prominent Ranger of that era named Gen. John Hunt Morgan. Morgan and his Confederate raiders raised havoc on the Union supply trains in Kentucky and Tennessee during the early part of the war. At a strength of 2,400 men built on recruiting on the move, Morgan's Rangers began a series of raids leading to what was termed the Invasion of Indiana where the whole countryside had taken up arms against them.[35] Near the end of July, in the vicinity of Liverpool, Ohio, Morgan was forced to surrender.

Equally skilled were Rangers under the command of Turner Ashby. Ashby achieved prominence as Thomas J. "Stonewall" Jackson's cavalry commander before he was killed in battle in 1862. He organized a cavalry company known as the Mountain Rangers. The Mountain Rangers assisted those sympathetic with the Confederates to pass into Virginia to disrupt railroad traffic on the Baltimore and Ohio Railroad and to interfere with the passage of boats on the Chesapeake and Ohio Canal.[36]

Once the Civil War ended, Rangers continued to serve as a guard force in the Texan frontier. In 1881, the Rangers fought their last battle with Indians, thus changing their primary mission from Indian defense to law enforcement. Texas maintained military Rangers almost continuously for the half a century and their legacy lives on as the Texas Ranger Division; a major division within the Texas Department of Public Safety.

WORLD WAR II RANGERS

Nearly 75 years elapsed without Ranger units amongst the ranks of the US Army. Emulating the techniques and procedures of the British Commandos, World War II saw the dawning of the modern Ranger and the modern Ranger organization when it activated six Ranger infantry battalions.

The very first commander of the 1st Ranger Battalion, Lt. Col. William O. Darby.

In May of 1942, Brigadier General Lucian K. Truscott recommended to US Army Chief of Staff, General George C. Marshall that an American unit be organized along similar lines as the British Commandos.[37] The initial intent was to train these men with the commandos, go on raids with them, and then be dispersed to other units to spread the gained knowledge. While this plan did not come to fruition, it bore a striking resemblance to the modern-day Abrahams Creed developed some 30-plus years later. Getting guidance from then-Maj. Gen. Dwight D. Eisenhower for this new organization, Truscott designated the new unit as the 1st Ranger Battalion in honor of Robert Rogers's Rangers.

William O. Darby was selected as the commander of this new organization. Darby activated the 1st Ranger Battalion on June 19, 1942, at Carrickfergus, Northern Ireland forming a force of 29 officers and 488 soldiers. Interestingly enough, Darby was the only Regular Army soldier in the entire command, so it can be said that the modern-day Ranger had its roots in the National Guard.[38]

Every member of the battalion was handpicked, and most were volunteers. Even though they were volunteers, it soon became apparent that some commanders had taken the opportunity to unburden themselves of malcontents in their ranks. Regardless, Darby took this eclectic group of "tough guys" and formed them into an organized, trained force.

Much of what the Rangers are today came from Darby's era. The ubiquitous Ranger Buddy System that permeates all aspects of the US Army today first found its way into the Ranger ranks during their train-up in Ireland in 1943. Darby ensured all Rangers worked together as teams, stipulating that the sum of the whole is better than any single man on the battlefield.

Today the buddy system is required in all Ranger and Regular Army units; working in pairs, taking care of the buddy to the left and right, and instilling teamwork in everything a soldier does and plans to do have become a way of life. The buddy system may be the first example of where Rangers and their training have benefited the entire army.

In support of the Canadians, Rangers led the way by being the first US troops in ground combat in Europe at Dieppe, France. Although the bulk of the forces committed to what was known as Operation Jubilee came from the Canadian Army, a small detachment of US Army Rangers were deployed consisting of 44 men and five officers under the command of Capt. Roy Murray.[39] The mission ultimately failed, however, the Rangers' role in the operation made front-page news in the United States. The *New York Post* led with the banner headline "We Land in France," while the *New York World Telegram* somewhat optimistically proclaimed that "US Troops Smash the French Coast."[40]

The official entry of the Rangers into the Second World War was in North Africa for Operation Torch, which began on November 8, 1942. Despite their uniqueness, the 1st Infantry Division Commander deployed the Rangers on conventional mission sets to support the 16th Infantry Regiment.[41] Following the landing and subsequent fights in North Africa, the Rangers did not see combat for another three months when they participated in an attack on Gafsa. There the Rangers were ordered to infiltrate and attack the enemy from the rear at a critical pass, supporting the 26th Infantry's attack from the front.

The success the 1st Ranger Battalion had in North Africa impressed the army's senior leaders, and Darby recommended to Eisenhower on April 14, 1943, to stand up the 3rd and 4th Ranger Battalions—an addition of 52 officers and another 1,000 men making up what was known as the Ranger Force.

SPEARHEADING ONTO SICILY AND NORMANDY: RANGERS LEAD THE WAY!

The 3rd and 4th Ranger Battalions were activated and trained by Col. Darby in Africa near the end of the Tunisian Campaign. The 1st, 3rd, and 4th Battalions continued to see themselves as the Ranger Force even though they were not authorized as a regimental headquarters to pull them together as an organized brigade. Regardless, the Rangers of this era found their identity and began the tradition of wearing the scroll shoulder sleeve insignia, which has been officially adopted by each of today's Ranger battalions as well as the Regimental Headquarters.[42]

Operation Husky, led by Darby's Rangers, consisted of the 1st, 3rd, and 4th Battalions. After just six weeks of training, the newly formed elite united spearheaded Patton's Seventh Army onto the island of Sicily on July 10, 1943. Known as Force X, the 1st and 4th Battalions were attached to Maj. Gen. Omar Bradley's II Corps—making an opposed landing at Gela where it captured the town and then defended it against a German and Italian counterattack.[43] It was here that Darby personally displayed heroism in the line of duty by borrowing an antitank gun and destroying a tank and was awarded the nation's second highest medal for heroism, the Distinguished Service Cross.[44]

Following the fight on Sicily the Allies launched an attack on the Italian mainland. Operation Shingle was a US Army VI Corps attack at the Anzio Beachhead. It was led by Lt. Gen John Lucas in an effort to unhinge the formidable German Gustav Line defenses to the southeast where Lt. Gen. Mark Clark's V Army had been fighting since January 16, 1944 at what was known as the first Battle of Monte Cassino. During this battle, the 1st, 3rd and 4th Ranger Battalions which formed what was known as 6615th Ranger Force supported the 3rd Infantry Division attack on Cisterna in an attempt to break out of the Anzio stronghold.

In command of the 3rd Infantry was Maj. Gen. Lucian Truscott who had been in Ireland supporting the formation of the Ranger units earlier in the war. At Anzio, Truscott met with Col. Darby and together they decided to have the Ranger Force infiltrate along the Pantano ditch, which runs northwest across the fields to the right of the Conca-Cisterna road to capture the town of Cisterna. This effort proved costly for the Rangers for the German presence was stronger than the intelligence reports had revealed and in the end out of the 767 Rangers who made it into Cisterna, only 6 made it out. The rest were killed or captured marking the end of the 6615th as an organized regiment.

Six months later the 2nd and 5th Ranger Battalions were involved in the D-Day landings at Omaha Beach, Normandy where the Rangers gained their now famous motto. The situation on Omaha Beach was becoming increasingly critical and drew concern from the assistant division commander of the 29th Infantry Division, Brig. Gen. Norman D. Cota, Cota "stated that the entire assault force must clear the beaches and advance inland or die."[45]

General Cota came down the beach. "In the Hollywood version, he calls out 'Rangers lead the way!' and off they charged."[46] It is doubtful that what, if anything, he actually said was heard, for the battlefield noise was so loud that he surely could not be heard even a few feet away. What we do know was that he moved from group to group.

One of the first people Cota encountered was Capt. Jack Raaen, the Headquarters company commander for the 5th Ranger Battalion. Cota recognized Raaen immediately because Cota's son was a West Point classmate of Raaen's son John. During this encounter, Raaen reported the location of the battalion commander's command post (CP) belonging to Lt. Col. Max Schneider (CP).[47]

As he moved to Schneider's CP, Cota started encouraging soldiers along his path to move out on their own, saying, "Don't die on the beaches, die up on the bluff if you have to die, but get off the beaches or you're sure to die."[48] To Raaen he said, "You men are Rangers and I know you won't let me down."[49]

Cota eventually located Schneider's CP, where "according to one witness, Cota said to Schneider, 'We're counting on you Rangers to lead the way.'" Sgt. Victor Fast, Schneider's interpreter, remembered Cota saying, "I'm expecting the Rangers to lead the way."[50] Other accounts had Cota arriving at the CP and asking Schneider, "What outfit is this?" Schneider supposedly answered, "5th Rangers Sir!" To which Cota replied, "Well god damn it, if you are Rangers get up there and lead the way!"[51]

Regardless of Cota's exact words, the motto of the Rangers became "Rangers lead the way." This motto has stood the test of time since World War II.

While the motto is inspiring, the Rangers of the day arguably had little need for the motivation that day. "There was little or no apprehension about going through the wire and up the hill," Corp. Gale Beccue of B Company, 5th Rangers, remembered.[52] "We had done that in training so many times that it was just a matter of course."[53]

Beccue and a private shoved a Bangalore torpedo under the barbed wire and blew gaps; they then proceeded to move on the German position, where they encountered little opposition since the Germans in their forward positions started pulling back to the rear. However, enemy artillery covered the German withdrawal and concentrated on the American Rangers occupying the former German locations. The implication of the Ranger motto that it took

Rangers to lead the Virginia National Guard unit, the 116th Infantry of the 29th Division, off the beach was faulty. In fact, the first organized company to the top at Vierville was Company C, 116th Regiment.[54] The truth was that the 116th preceded the Rangers off the beach. The members of the 116th still at the seawall came from those companies that had been decimated in the first wave.[55]

The 2nd and 5th Battalions were introduced on May 6, 1944, as two additional European theater Ranger Infantry Battalions and were named the Provisional Ranger Group. With Lt. Col. James Earl Rudder as their commander. The Provisional Ranger Group got "their baptism by fire on the beaches of Normandy on D-Day as part of Operations Neptune."[56]

The 2nd Battalion's mission was most daring of any of the D-Day assaulting units. It was to land about four miles west of the right flank of Omaha Beach on a narrow strip of beach at the base of Pointe du Hoc and scale a 90-120-foot sheer cliff under heavy enemy fire, cross the obstacle and mined crest, and destroy what appeared to be six 155mm cannon located in the open, reinforced concrete casemates.[57] The remainder of the Ranger force, to include the 5th Ranger Infantry Battalion, "would wait off shore under the command of Lieutenant Colonel Max Schneider and prepared to assist if needed.

As the Rangers scaled the cliffs, the Allied destroyers provided them with fire support and ensured that the German defenders could not fire down on the assaulting troops. The Rangers regrouped at the top of the cliffs, and a small patrol went off in search of the entrenched guns and destroyed them with thermite grenades. The assault on Pointe du Hoc has recently been portrayed in a series of video games titled *Call of Duty 2, G.I. Combat* and *Company of Heroes.*[58] The movie *The Longest Day* also contains scenes of the assault on the cliffs of Pointe du Hoc.

After Pointe du Hoc, the 2nd Ranger Battalion pushed westward to clear the battlefield of enemy pockets of resistance located along the coastline of the Cherbourg Peninsula. Both the 2nd and 5th Ranger Battalions were used as conventional infantry to help reduce the defenses of the port city of Brest in September 1944 and then eventually on to the Hurtgen Forest. Here they were alerted on December 6th and attached to the 78th Infantry Division, defending the left flank of the "Battle of the Bulge" commencing on December 16, 1944.

Near the end of March 1945, the 2nd Ranger Battalion crossed the Rhine and was pressed back into more combat. Firefights and skirmishes were few and far between at this point in time, and the men engaged in the mop-up reducing what was left of German resistance. More attention was now paid to snipers and saboteurs hiding behind the enemy lines. In the first week of May 1945 the battalion was suddenly moved to Czechoslovakia, where further skirmishes were encountered and neutralized.[59]

After the breakthrough on the beach, the 5th Rangers spent a few weeks training and then relieved elements of the 2nd Infantry Division northwest of Brest. In the fall of 1944, the 5th Rangers were attached to the 6th Cavalry Group, part of General Patton's battlefield eyes and ears. During the Battle of the Bulge, the Rangers' mission was a defensive one at St. Avold where the German attack never materialized. The battalion then moved back into a training mode in early February and was attached to the 94th Division. They eventually took part in the bloody Irsch-Zerf Campaign finishing the final two months of the war fighting as conventional infantry conducting routine combat missions, guarding prisoners, and imposing military governments in and around Bamburg, Erfurt, Jena, Gotha, and Weimer.[60]

RANGERS IN THE PACIFIC

While the Rangers in Europe were conducting one of the most daring raids in a long list of Ranger accomplishments, in the Pacific theater the 6th Ranger Battalion was preparing for their own daring mission. In contrast to the Rangers fighting in a mostly conventional role on the African and European continents, the 6th Ranger Battalion, operating in the Pacific, was employed against tasks for which it was specifically organized and trained to include reconnaissance, infiltration operations behind enemy lines and direct action attacks.

Converting from the 98th Field Artillery Battalion in New Guinea to the 6th Ranger Battalion under the command of Lt. Col. Henry "Hank" Mucci was no easy task. Mucci informed the battalion it was being converted to Rangers and would return to the Philippines causing a large turnover of personnel. Intense training ensured and on October 1944, the battalion deployed on three APDs for Leyte in the Philippines.[61] Here they rose the first American flag on Philippine soil as part of General Douglas MacArthur's "Return to the Philippines."

What ensued next was one of the most successful rescue operation in the history of the US Military and many books and movies have captured the raid in detail. To conduct this mission, Lt. Col. Mucci selected Capt. Robert Prince to lead the raid on Cabanatuan; the prison camp housing Americans who survived the famed Bataan Death March. The Rangers hit the camp in a daring night raid and brought out 512 prisoners of war, killed about 200 enemy troops, and only losing two Rangers.

Through the remainder of the war the battalion operated as individual companies getting into a serious of skirmishes. With the attack on Aparri combat had ended in the Philippines for the 6th Rangers, and they began to prepare for the invasion of Japan. When the Atomic bomb was dropped on Japan on August 6, 1945 the Rangers were sent to Japan as occupation forces. On December 30, 1945, the unit was deactivated in Japan and the individual Rangers were either sent home or assigned to other units.[62]

MERRILL'S MARAUDERS

During the Second World War, there was another unit besides the six battalions officially designated as Ranger organizations that clearly operated as a Ranger force. In September 1943, the US formed a long-range-capable unit that could penetrate deep into enemy territory. This unit was the 5037th Composite Unit (Provisional), code-named "Galahad" and soon became known as Merrill's Marauders in honor of its eventual commander, Brig. Gen. Frank D. Merrill. And while the 5037th may not have been an official Ranger organization, its legacy may have had more impact on the present-day Rangers than any other official Ranger unit of that era.

The Marauders ended up operating in the China-Burma-India (CBI) Theater where the mountainous, rain-infested, forested terrain of Burma where the temperature, humidity, and mosquitoes were stifling to any force unfortunate enough to operate in that theater. This three-battalion-sized regiment was known for their ability to negotiate the difficult Burmese terrain with apparent ease. Fighting under the overall command of Lt. Gen. Joseph W. "Vinegar Joe" Stilwell, the commander of Galahad, Merrill's Marauders conducted a series of attacks against Japanese supply lines, units, and lines of communication (LOC).

Colonel Frank D. Merrill, Commander of the 5037th Composite Unit (Provisional), better known as Merrill's Marauders.

The goal for this unit was to support the building of the Ledo Road, which was vital to the survival of the Burmese nation in its battle with the Japanese occupiers. In five major and 30 minor engagements, the Marauders marched over 1,000 miles through extremely difficult and often unnavigable territory against the Japanese 18th Division, famous for their conquering exploits in Singapore and Malaya. The Marauders eventually attacked and seized key terrain at the Myitkyina Airfield in May 1944. Completely exhausted and suffering over 50% casualties either by enemy fire or environmental inflictions, the 5037th was relieved on August 10th and consolidated with the 475th Infantry. The 475th eventually became the 75th Infantry on June 21, 1954, and even though it would be inactivated on March 21, 1956, it became the forefather of today's 75th Ranger Regiment.

RANGERS IN KOREA

The modern Rangers owe much to their history to the Korean War–era Rangers, better known then as Rangers Infantry Company (Airborne) or RICA. While the combat experience and courage displayed on the battlefield are what helped shape the modern force, other facets of that era have had an equal impact. Today's Ranger Training Brigade (RTB)—the unit responsible for training soldiers and turning them into "tabbed" Rangers—had its roots during this period with the establishment of the Ranger Training Center.

The outbreak of hostilities in Korea in June 1950 once again signaled the need for the skill sets and organizational makeup of a Ranger Force. And while the war lasted until 1953, much of the Ranger involvement was over by 1951.

The United Nations (UN) forces organized in the defense of the Pusan perimeter had little means of gathering intelligence as the North Korean force surrounded them. The North Koreans had the capabilities to infiltrate small teams into the South and the Americans decided it needed this type of capability to patrol a small salient north of the Pusan Perimeter. Identifying this need, the Joint Chiefs of Staff called for a "Marauding Company" to be formed, taking the name from Merrill's Marauders' fame in World War II. The Marauding name never took hold, and the decision was made once again to form Ranger units.

Col. John H. McGee, was given the daunting task of organizing and training a commando-style unit. One of the men McGee picked to command the Rangers—and one of the most famous, successful, and loved Rangers today—came from the ranks of Korean War Ranger heroes: honorary Colonel of the Regiment Ralph Puckett.

Led by then-2nd Lt. Puckett, the Eighth Army Ranger Company was formed in August 1950. It would serve as the role model for the rest of the Ranger units in Korea to follow. Puckett volunteered to be a Ranger, for he "wanted to be the best."[63] He ended up leading the Eighth Army Ranger Company during the Korean War and was awarded the Distinguished Service Cross (DSC) for his actions on November 25, 1950, when his company of 51 Rangers was attacked by several hundred Chinese forces at the battle for Hill 205. Retiring in 1971 Puckett was eventually appointed as the Honorary Colonel of the 75th Ranger Regiment on July 19, 1996 and is revered by every Ranger in the regiment today.

The Korean War is replete with examples of heroes from the RICA, and Ralph Puckett is one of them. Formally designated as the 8213th Army Unit (AU), the Eighth Army Ranger Company (8ARC) was attached to the 25th Infantry Division, IX Corps, and ordered into action on October 10. It participated in the drive to the Yalu and was deactivated in March 1951. What this experience ultimately provided for the Rangers was a lesson in how to form and organize a force quickly for a combat need.

Officers at Fort Benning had long studied the employment of Ranger units. They recognized that the organization of Ranger infantry battalions offered many advantages, including better tactical employment. They believed that a battalion commanded by a lieutenant colonel could operate more effectively with the senior officers of a division or high level staff, than could a captain who commanded a ranger company working for a two-star command. Additionally the addition of a ranger battalion staff would provide the oversight needed for the welfare of the men.[64]

The Korean War effort initially called for one Ranger company per each division in the army. The new Ranger Training Center argued this was inefficient but conceded and formed sixteen additional Ranger companies in seven months.[65] The plan was to send men from units through the school and back to the unit – much like the Army does today when qualifying Rangers for the Ranger Tab. Thus, the search for Ranger volunteers from the Big Army ensued.

In the 82nd Airborne Division, the call for volunteers to join the Rangers was astounding. Some estimates were as high as 5,000 experienced Regular Army paratroopers took interest. Given this massive response, the ruthless sorting out process began. Where possible, selection of the men was accomplished by the officers who would command the companies, similar to colonial days when Robert Rogers was recruiting.

Orders were issued and those selected were shipped to Fort Benning, Georgia.[66] All Ranger volunteers were professional soldiers. Some of the men had fought with the original Ranger Battalions, the First Special Service Force, or the Office of Strategic Services during World

War II. Many of the instructors were drawn from the same group. These were men highly trained and experienced in Ranger operations during World War II.

The training was extremely rigorous. It consisted of amphibious and airborne operations including low-level night jumps, demolitions, sabotage, close combat, and the use of maps. Every American small-arm, as well as those used by the enemy, was mastered. Communications, as well as the control of artillery, naval, and aerial fires, were stressed. Much of the training was at night.

Physical conditioning and foot marching were constant. The goal was to prepare a company to move from 40-50 miles, cross-country, in 12-18 hours. Men learned that they could withstand sleep deprivation, the lack of food and the exhausting environmental hazards often found in combat. No man was forced to remain a Ranger candidate. During training, there was a jeep in the background with a white flag. Anyone who decided he did not want to, or could not, continue had only to go sit in the jeep. No one would harass or mock him. He would be driven away and his personal gear removed from the barracks before the other men returned.

Nowhere in American military history is the volunteer spirit better expressed then it was with Ranger volunteers in Korea. They were volunteers for the Army, for Airborne training, for the Rangers, and for combat. At a time when United Nations forces numbered over a half a million men, fewer than 700 were Airborne Rangers.

August 1951 brought the end of the Rangers in Korea and the inactivation of the rest of Ranger Infantry Companies. The focus now was on the development of a new program to make all airborne, infantry, and armored units capable of Ranger-type missions. To do this, the Army developed a new Ranger program at the Infantry School at Fort Benning where the production of qualified small-unit leaders was the goal; arguably the birth of the modern day Ranger School.

THE VIETNAM WAR RANGERS

In the waning years of the Cold War there was a universal need for small units to conduct deep penetration intelligence gathering missions. From this need was born the Long Range Patrol (LRP) or Long Range Reconnaissance Patrol (LRRP). Concealing themselves from the enemy eyes by infiltrating enemy lines any means, these teams were intended to report enemy movements and provide early warning. With a focus on stealthy operations instead of direct action the LRP/LRRPs would eventually form the basis for the Ranger companies of Vietnam.

Prior to the establishment of these units, individual Rangers were called to duty to employ special skills to the South Vietnamese special operations forces. These South Vietnamese forces were counter guerilla teams called Biet-Dong-Quan (BDQ) and a significant number of US Ranger-qualified officers and NCOs served as their advisors. Graduation from the South Vietnamese Ranger type course was a "real world" combat operation. Those who returned alive were graduates. Overall there were a total of eight BDQ regimental sized groups with 22 battalions – each assigned five US Ranger-qualified officers and NCOs as advisors.[67]

With the growing United States involvement in the Vietnam, LRP units were officially authorized and infused into the Army formations. Eventually, the mission designator of "Reconnaissance" was dropped and became solely LRP units as they performed not only reconnaissance type missions but also combat missions such as ambush, prisoner snatch and raids. As a result, all major commands activated a 118 man LRP company with both the 20th Infantry Regiment and the 51st Infantry Regiment establishing 230 man companies. Separate brigades organizing their LRP companies into 61 men.

Long Range Reconnaissance Patrol units operating throughout the four Military Regions of the Republic of Vietnam:[68]

Unit Major Command
Co. D, 17th Infantry, (LRP) V Corps Federal Republic of Germany
Co. C, 58th Infantry (LRP) VII Corps Federal Republic of Germany
Co. E, 20th Infantry (LRP) I Field Force Vietnam
Co. F, 51st Infantry (LRP) II Field Force Vietnam
Co. D, 151st Infantry (LRP) II Field Force Vietnam
Co. E, 50th Infantry (LRP) 9th Infantry Division
Co. F, 50th Infantry (LRP) 25th Infantry Division
Co. E, 51st Infantry (LRP) 23rd Infantry Division
Co. E, 52d Infantry (LRP) 1st Cavalry Division
Co. F, 52nd Infantry (LRP) 1st Infantry Division
Co. E, 58th Infantry (LRP) 4th Infantry Division
Co. F, 58th Infantry (LRP) 101st Airborne Division
71st Infantry Detachment (LRP) 199th Infantry Brigade

Captain Dave Grange with his Platoon Sergeant in Vietnam as Rangers in Company L, 101st Airborne Division. (Source: Grange)

74th Infantry Detachment (LRP) 173rd Airborne Brigade
78th Infantry Detachment (LRP) 3rd Brigade, 82nd Airborne Division
79th Infantry Detachment (LRP) 1st Brigade, 5th Mechanized Division

By February 1, 1969, most of the active army LRP companies and detachments were deactivated as LRP units and reactivated as fifteen separate companies of the 75th Infantry Regiment. Thirteen of these companies served in Vietnam until the entire force and was inactivated once again on August 15, 1972.

The following is a list of Ranger units that served during this era. While the lineage of today's Rangers aren't officially from these units it is instructive to list the Ranger units of this era:[69]

Unit Major Command / Period of Service
Co A (Ranger), 75th Infantry Ft Benning / Ft Hood 1 Feb. 1969 – 15 Oct. 1974
Co B (Ranger), 75th Infantry Ft Carson / Ft Lewis 1 Feb. 1969 – 15 Oct. 1974
Co C (Ranger), 75th Infantry I Field Force Vietnam 1 Feb. 1969 – 25 Oct. 1971
Co D (Ranger), 151st Infantry II Field Force Vietnam 1 Feb. 1969 – 20 Nov. 1969
Co D (Ranger), 75th Infantry II Field Force Vietnam 20 Nov. 1969 – 10 Apr. 1970
Co E (Ranger), 75th Infantry 9th Infantry Division 1 Feb. 1969 – 12 Oct. 1970
Co F (Ranger), 75th Infantry 25th Infantry Division 1 Feb. 1969 – 15 Mar 1971
Co G (Ranger), 75th Infantry 23rd Infantry Division 1 Feb. 1969 – 1 Oct. 1971
Co H (Ranger), 75th Infantry 1st Cavalry Division 1 Feb. 1969 – 15 Aug. 1972
Co I (Ranger), 75th Infantry 1st Infantry Division 1 Feb. 1969 – 7 Apr. 1970
Co K (Ranger), 75th Infantry 4th Infantry Division 1 Feb. 1969 – 10 Dec. 1970
Co L (Ranger), 75th Infantry 1O1st Airmobile Division 1 Feb. 1969 – 25 Dec. 1971
Co M (Ranger), 75th Infantry 199th Infantry Brigade 1 Feb. 1969 – 12 Oct. 1970
Co N (Ranger), 75th Infantry 173rd Airborne Brigade 1 Feb. 1969 – 25 Aug. 1971
Co 0 (Ranger), 75th Infantry 3rd Brigade, 82nd Abn. 1 Feb. 1969 – Division 20 Nov. 1969
Co P (Ranger), 75th Infantry 1st Brigade, 5th Mech. 1 Feb. 1969 – Division 31 Aug. 1971

The Rangers in Vietnam were volunteers like they were in any other era. Training was a combat mission for these volunteers. Volunteers were assigned and not accepted in the various Ranger companies until after a series of patrols by which the volunteer had passed the acid test of a Ranger – combat- and was accepted by his peers. Following peer acceptance, the volunteer was allowed to wear the black beret and red, white and black scroll shoulder sleeve insignia bearing his Ranger company identity.[70] The successes achieved led to multiple awards to include the following campaign streamers that hang proudly on the 75th Ranger Regiment colors today:

75th Infantry (Ranger) in Vietnam Campaign Streamers and Unit Awards
Campaign Streamers, Vietnam
Counteroffensive Phase VI
Tet 69 Counteroffensive

Summer–Fall 1969
Winter- Spring 1969
Sanctuary Counteroffensive
Counteroffensive Phase VII
Consolidation I
Consolidation II
Cease Fire
Decorations, Vietnam
RVN Gallantry Cross w/Palm – 23 Awards
RVN Civil Actions Honor Medal – 10 Awards
US Valorous Unit Award – 6 Awards
US Meritorious Unit Commendation – 2 Awards[71]

When the war in Vietnam ended in 1974 the US Army was in disarray in many ways. Discipline issues, a drug pandemic, racist struggles, dissention, and even desertion all had threads within the ranks. The unpopularity of the Vietnam War with the American population did not bode well with the Army itself and the Soldiers – who were mostly conscripts – did not conform to a less than savory way of life. Soldiering in the US of A was at – perhaps – its lowest point. Something had to be done and the leadership knew this challenge would be worth making some bold changes in order to survive as an Army.

2

The State of the Army in 1974

"Discipline is the soul of an army. It makes small numbers formidable; procures success to the weak, and esteem to all."

George Washington
American Commander in Chief of Colonial Armies
in the American Revolution (1775–1783)

DISILLUSIONMENT AND ILL DISCIPLINE

The US participation in the war in Vietnam left the US military a weak and practically impotent institution. The public opinion and trust in the Army was at a low as the war in Vietnam drew to an end. Many Americans blamed the military for the war as much as they blamed the civilian policymakers.[1] Returnees feeling that they served with honor faced a hostile, or at best, an indifferent public reception. The reactions toward the returning veterans ran the spectrum from discontent to distrust and, more times than not, total disdain.

That type of reception, coupled with, in many cases, horrific combat experiences, left many returning soldiers lost and looking for less than optimal ways to cope with issues. Unlike the past 25 years, where the American public generally supported US forces that deployed to Desert Storm, the small conflicts in the 1990s and the full out wars in Iraq and Afghanistan, the Vietnam soldiers could feel the public's contempt. In order to cope, a number of soldiers became drug addicts while at war, where the supply of drugs like heroin was plentiful. Upon returning, this drug infestation festered and grew, causing obvious issues in morale and discipline in the ranks.

Kent Woods in his War College Strategy paper writes about how a former cadet, who was a tactical officer at West Point, relayed to him his experience while on his Cadet Troop Leading Training (CTLT) assignment in the 1970s. During CTLT, cadets are typically assigned in their third year at West Point to an army unit as an acting platoon leader for about six weeks in the summer before their senior year.[2]

This cadet was assigned to a mechanized unit at Fort Riley, Kansas. While on a field exercise, the soldiers of the platoon were having difficulty maneuvering their M113 Armored Personnel Carriers (APC). The cadet spent the entire field exercise constantly cajoling, yelling, wheedling, and pleading to get them to perform tactically. Then it happened. The platoon pulled into yet another defensive position, and, as he puts it, "with alacrity the soldiers began cutting down the fauna in the area to place it on top of the APCs as camouflage. The cadet was initially proud that his leadership abilities had finally [borne] fruit and the platoon was doing things energetically correct. Then he realized that the local fauna was in fact marijuana. The soldiers were stacking it on top of the exhaust vents of the APCs to dry it out."[3] This an example of how the Army and its soldiers were in the post-Vietnam era.

The Chief of Staff of the Army, Creighton Abrams, understood the extent of the drug problem in his force. Found in his personal papers are the notes from staff meetings he attended. One particular note, dated February 26, 1973, was an annotation made regarding the problems in the Army in Europe specifically:

Bad Drugs hard drugs in Frankfurt
Hash available replaced by hard
Age 25 and under 50% to 90%
Soldiers bored
VOLAR term no good
Permissiveness, discipline
Chain of Command.[4]

It isn't possible of course to know what Abrams was thinking when he wrote these notes. However, it is obvious that Abrams was not only aware, but clearly concerned about the state of the Army in the post-Vietnam era.

On February 22, 1974, Abrams testified to the Senate Armed Services Committee, or SASC. A good portion of his testimony dealt with the drug problem in the United States Army in Europe, or USAEUR, as well as the rest of the US Army. In his testimony, Abrams stated that the drug problems in the Army in 1974 matched the degree of the problem that existed during the Vietnam War and having a great impact on the readiness of the force. Senator Harold Hughes present that day, as a member of the SASC made the following statement to the press about Abrams: "He sees a very serious question ... I asked him if it was comparable to Vietnam, and he said yes."[5]

Drugs and disillusionment with leaders and the mission caused a breakdown in discipline. As it became apparent that the US was pulling all the strings it could to get out of Vietnam, discipline, especially in the rear base camps, began breaking down in many units toward the end of the war. No one wanted to be the last man to die in Vietnam. The focus to just make it home alive resulted in a variety of malcontent soldiers resulting in number of leadership challenges to include racial tension and even instances of 'fragging' (tossing a fragmentation grenade into the sleeping quarters or office of a superior officer or noncommissioned officer to injure or 'warn') causing obvious health and morale problems in military organizations.[6]

Col. (retired) Jerry Barnhill, original company commander for C Company, 1st Ranger Battalion, remembers from his second tour in Vietnam "how sick the Army was." As the battalion personnel officer, or S1, he recalls being shocked when the battalion commander was relieved for ordering a soldier to go to the field. It certainly was a leadership challenge that made it difficult to command a cohesive, tight-knit outfit when those who refused to conform were being pandered to while those who followed the code of conduct were willingly going to the field and dying on the battlefield. According to Col. (retired) Gerry Cummins, who served in the Rangers from 1978 to 1981, the sine wave of proficiency to deficiency seems to have been the unfortunate end result of an unpopular war in the 1970s. The true professionals were few and far between.

The Army fresh out of Vietnam was struggling. These difficulties were notably apparent in the overseas garrisons in Germany and Korea. Drug problems, poor leadership especially

COL Ronald F. Rokosz (Commander, 2d Brigade, 82d Airborne Division; without helmet) confers with liaison team from the French 6th Light Armored Division in tactical assembly positions near Rafha in the Northern Province of Saudi Arabia on 23 February 1991, the day before the ground offensive for Operation Desert Storm began. (XVIII Airborne Corps photograph DS-F-317-11A by MAJ Dennis P. Levin)

Brig. Gen. Ronald F. Rokosz; original member of the 1/75th Rangers and seen here in his command photograph as the Commanding General, Special Operations Command, Pacific (SOCPAC). (Courtesy of Ronald F. Rokosz).

at the junior NCO and officer levels, and severe racial problems divided the army at multiple levels. As with the times, race riots were not uncommon, especially in the understrength caserns of Germany as the Army tried to rebuild its European combat power.

Brig. Gen. (retired) Ron Rokosz, original commander of B Company, 1st Ranger Battalion, has the distinction of being the only soldier to command three Ranger battalions and serving in the Regimental Headquarters in various positions to include the deputy commander for the late Gen. (retired) Wayne Downing, who was the regimental commander. Gen. Rokosz recalls that Vietnam had an impact not only on those fighting in Southeast Asia and on those in the United States, but also on the military communities abroad. He tells of his tour as a platoon leader in Augsburg, Germany, where there were only 10 officers in the entire battalion mainly because the officers assigned to Germany and other European countries did short tours in Europe and then were sent to Vietnam for a year. Gen. Rokosz goes on to explain that later, while serving with the U.S.-based 1st Cavalry in Vietnam, he learned that the army was not a well-knit organization like it is today; in fact, there were places on forward-deployed firebases where the officers just wouldn't go after dark.

Rokosz remembers distinctly that two soldiers from his unit in the 1st Air Cavalry Division refused to move out with their company. In most cases like this the commanders' hands were tied as to what they could do. Most just placed them in different parts of the camps doing administrative duties or oftentimes sent them to the larger camps in the "rear."

But it wasn't just casual refusals to follow orders. The US Army in Vietnam witnessed actual acts of mutiny with the very first mutinies recorded in 1968 as platoon-level refusals to fight occurred.[7] The army recorded 68 such mutinies that year. By 1970, in the 1st Air

Cavalry Division alone, there were 35 acts of combat refusal.[8] One military study concluded that combat refusal was "unlike mutinous outbreaks of the past, which were usually sporadic, short-lived events. The progressive unwillingness of American soldiers to fight to the point of open disobedience took place over a four-year period from 1968–71."[9]

By the end of the year the combat refusals of individuals and small units had expanded to whole companies. The first reported mass mutiny was in the 196th Light Brigade in August 1969. Down to 60 men from its original 150, Company A of the 3rd Battalion, was operating in Songchang Valley. Under heavy fire for five days it eventually refused an order to advance.[10] Word of the mutiny spread rapidly. The New York *Daily News* ran a banner headline, "Sir, My Men Refuse to Go."[11]

The lack of punishment for mutinous actions led to other individual and unit's refusal to follow orders for the lack of repercussion left them fearless. Hanoi's *Vietnam Courier* documented 15 different GI rebellions in 1969. At home, the *CBS Evening News* broadcast live a patrol from the 7th Cavalry telling their captain that his order for direct advance against the NLF was nonsense, that it would threaten casualties, and that they would not obey it. Another CBS broadcast televised the mutiny of a rifle company of the 1st Air Cavalry Division.[12]

When Cambodia was invaded in 1970, soldiers from Fire Base Washington conducted a sit-in. They told *Up Against the Bulkhead*, "We have no business there … we just sat down. Then they promised us we wouldn't have to go to Cambodia."[13] This unaddressed precedence was followed by two additional mutinies, as men from the 4th and 8th Infantry refused to board helicopters to Cambodia.[14] During the March 1971 invasion of the Vietnam border country Laos, two platoons refused to cross the border. To prevent a spreading mutiny, the entire organization was pulled out of the Laos operation. As a result, the captain was relieved of his command, but there was no discipline against the men.[15]

As this type of environment evolved, the introduction of conscientious observer (CO) status as an appropriate and acceptable reason to refuse to serve came in a mad rush by thousands attempting to qualify based on religious beliefs not to go to war. While this author agrees that this choice is a right, if not noble, its use as a means of getting out of going to war exacerbated the already growing disciplinary problems in the active force. Not registering for the draft based on CO or other statuses quickly became an issue. During the Vietnam War, at least 250,000 men refused to register. Only 250 of them were ever convicted in court. With a man like Muhammad Ali, the world prizefighter who qualified for CO status, becoming a model for those against the war, the popularity of this status soared.[16]

There is no doubt that the Vietnam War was an unpopular war. Historical precedents do not exist for some of the services' problems, such as desertion, drugs, mutiny, unpopularity, seditious attacks, and racial troubles. Nowhere, in the history of the Armed Forces have comparable past troubles presented themselves in such general magnitude. Something had to be done and the leadership knew this challenge would be worth making some bold changes in order to survive as an Army.[17]

Many of the problems within the ranks stemmed from the lack of standards and leadership and the leadership at the time instigated an Army wide study lead by the United States Army War College (USAWC). Gen. William Westmoreland, then the Army Chief of Staff, directed the War College to conduct a study. This review was considered a seminal work that reviewed the military professionalism of the officer corps as a means to better understand the challenges. The study was conducted by a group of officers who were students at the War College and included such future leaders of the army as Lt. Gen. Walter F. Ulmer Jr., who was destined to command III Corps. The group conducting the study interviewed 450 officers ranging in rank from lieutenant to colonel. Comments collected by the study provide a window into what the officers of the time were feeling.[18] One problem recorded by the study was the lack of standards: "The lack of uniform standards throughout the Army … standards of appearance and standards of performance. Problems that every commander is faced with today… the haircut; on every single post and on each post, within units, there is a different standard for haircuts … what they would like is a Department of the Army standard that is enforced by all commanders and all commanders have to live with it …"[19]

One item in the study was a review of haircut standards. While these standards seem like a minor point of contention in today's Army they clearly were on the mind of officer as representative of their issues in 1970.[20] Today such standards, thanks to the inclusion of the Ranger formations in our army as a change agent for the entire force, endure as touchstones of a unit's discipline and morale.

Several other problems arose from this review. Of particular note is the following conclusion by the panel:

> The most frequently recurring specific themes describing the variance between ideal and actual standards of behavior in the Officer Corps include: selfish, promotion oriented behavior; inadequate communication between junior and senior; distorted or dishonest reporting of status, statistics, or officer efficiency; technical or managerial incompetence; disregard for principles but total respect for accomplishing even the trivial mission with zero defects; disloyalty to subordinates; senior officers setting poor standards of ethical/professional behavior.[21]

VAGUE MISSIONS

Gen. Volney Warner, the eventual Commander in Chief, Readiness Command (REDCOM), headquartered at MacDill Air Force Base in Tampa, Florida, and the assistant division commander of the 82nd Airborne Division and eventual deputy chief of staff for operations at Forces Command at the time of the ending of the Vietnam War inferred in an August 7, 2010, interview that the state of the Army reflected the state of the country at the time.

Maj. Gen. K. C. Leuer, the eventual first commander of the 1st Ranger Battalion, said in a 2010 interview that "the Army was at its lowest point in history; it had lost its focus." Leuer said that the army's state was a reflection of what was going on in the nation—turmoil over integration, lack of discipline, absences without leave (AWOL), and insubordination all became prevalent and had a severe impact on an already weakened NCO corps. He claimed that "shake and bake" NCOs (those noncommissioned offers who were mass-produced to support Vietnam) were unable to provide leadership and discipline at the unit level.

However, there was some dissension in how the Ranger should be infused into the Army. Volney believed that the eventual formation of a Regiment of Rangers was a big mistake. He believed instead that each conventional division should have its own Ranger battalion where the NCOs could be the standard bearers for the entire force.

Col. (retired) Jerry Barnhill had his own concerns about mission sets assigned to the Ranger Battalion. Unlike Volney and Leuer, he felt the Rangers should not be supporting the conventional army but should move toward a more special operations focus. The reader will

Maj. Gen. KC Leuer, Commander The US Army Infantry School and Fort Benning, Georgia circa 1988-1989. (Source: Leuer)

see later, as the Regiment came onboard and the mission sets evolved, that the Ranger units, while maintaining that they were the best light infantry in the world, tended to be more special operations oriented. Regardless of their orientation, the establishment of a Ranger Battalion, if for no other reason than to set the standards, had to happen.

Leuer was disgusted coming out of Vietnam at what was happening to the army in what many termed a quagmire when defining the Sisyphean nature of the war in Southeast Vietnam. The "search and destroy" mission was degrading, according to Leuer, who asked: "to search where? to find what?" He believed there was no clear mission or way of measuring success; neither were ever really defined in any of the battles on the operational and tactical levels.

The army was desperate for a clear set of performance measures. Criteria for success, definitive tasks under prescribed conditions, and standardized measures are what any army needs to perform, and the US Army in Vietnam lacked this requirement. Just like a new Ranger Battalion establishing a well-disciplined, capable force, the establishment of performance standards would set in motion its success.

With the need for what Leuer called performance-oriented standards, or POT, came an added monumental task of shifting from a conscription force to an all-volunteer force. Having just come out of Vietnam, the US decided to shift the military to an all-volunteer system. To Leuer, the army at this time was at the lowest ebb it could be and still hang together. The focus of what the army was all about, according to Leuer, had been lost; certainly it had little or no real professionalism left it except for what some units might put together.

With the expiration of Selective Service induction authority on June 30, 1973, the establishment of a new, all-volunteer army was under way. Many wondered if the army could recover sufficiently to recruit enough quality soldiers, and even if it did, many doubted the country would be able to pay the bill. The result was far from certain.[22]

To Leuer, integration played a large part in the challenges the army was tackling. The United States was in a state of turmoil with integration and all that it entailed. Leuer recalls that in 1967, while preparing the 3rd Battalion, 503rd Airborne Battalion, from Fort Campbell for a combat assignment with the 173rd Airborne Brigade, they were suddenly diverted to Detroit for the riots. The 1967 Detroit riot, also known as the 12th Street Riots, began on July 23, 1967. The precipitating event for the riots was a police raid on an unlicensed after-hours bar known as the Blind Pig located on the corner of Clairmont Street and 12th Street, which today is fittingly named Rosa Parks Boulevard.

The Blind Pig was in a predominantly black neighborhood where police expected to find a few patrons but instead found 82 people inside holding a party for two returning Vietnam veterans. President Johnson, in a somewhat unprecedented move, sent in federal forces, and Governor George Romney ordered in the Michigan National Guard resulting in dozens dead, hundreds injured and thousands incarcerated.[23]

This was one event in a time of disorder where riots were breaking out everywhere, sometimes inspired by the Vietnam experience and sometimes by integration *and* Vietnam. What was difficult for military leaders at the time was how these troubles manifested in the army—not only in terms of the soldiers as they emulated societal norms but also with the leaders, who also had the heavy responsibility of trying to find a way to gain good order and discipline.

Reliance on the noncommissioned officer (NCO) corps had always been the default in any army when discipline problems existed. The NCO corps prior to the Vietnam War was, according to Leuer and others, "strong" but only in relative terms. Today the noncommissioned officer education system or NCOES produces highly qualified standard-bearers for the NCO Corps.[24]

During the Vietnam era, the army established the Noncommissioned Officer Candidate Course (NCOC), a three-month school that promoted an E-2 or E-3 graduate into a SGT (Sergeant) E-5. When Msg. Sgt. (retired) Don Feeney was just 17 years old and a new private to the army, he was sent to the three-month course and was made an E-5 sergeant, but wasn't old enough to even deploy to Vietnam. Feeney eventually became a world-class Ranger. However, there is no doubt that in an attempt to raise the number of NCOs available to the fighting force, the risk associated with quickly making young soldiers into NCOs with no experience in the fight caused many problems. The NCO corps in Vietnam, while it had moments of brilliance, was struggling. The "shake-and-bake" NCOs had little to no leadership or combat experience. Putting these green young leaders in charge of small formations of disillusioned junior soldiers oftentimes led to great problems. In reality, leaders from NCO to senior officer at the time varied from very to not so professional to totally incompetent, much like they do in any era. Unfortunately, a large percentage of this population found itself in the bottom bin of this spectrum.

With a level of discipline at an all-time low and AWOLs and court-martials at an all-time high it was not uncommon for buses to go around to local jails near US posts to pick up troubled troops. On most posts the detention facilities were maxed out. All in all, Leuer thought, like many leaders throughout the army, that the Great Green Wave lost its focus and didn't know where it was going.

Gen. William R. Richardson, the assistant commander of Fort Benning and commandant of the Infantry Center at the time of the forming of the Rangers, said that in 1974 the army was in a state of change and trying to find its way. According to Richardson, the army needed reorganization, refocusing, and a drastic revamping of its training regimen. This situation was exacerbated by serious problems with drugs and race relations throughout the force.

In the Arab-Israeli Yom Kippur War, which ended in 1973, the great powers of the US and the Soviet Union were at odds. At the time it looked as though the US would have to prepare to rapidly deploy a force in reaction to multiple military requirements. The Pentagon grew concerned about its strategic ability to quickly move well-trained infantry forces to any spot in the world. Even though the 82nd Airborne Division was considered "light," it still required a tremendous amount of airlift capability to get it to the fight. The army leadership knew that something had to be done. Not only could the army not maintain a disciplined force, but also it did not have the capability to rapidly deploy and fight as both a special operations force and an elite infantry fighting organization. But first something had to be done to fix the challenges presented by the lack of performance standards in training.

PERFORMANCE ORIENTED TRAINING (POT)

Things began to turn when General Abrams selected Gen. William DePuy to form the Army Training and Doctrine Command (TRADOC) in 1973. DePuy, who was deeply affected by his experiences in World War II and the poor training of many of the units that were fielded in that war and the wars thereafter that led to needless deaths, began immediately to establish new training parameters and doctrine.

As a senior leader in Vietnam, DePuy would go out on patrols with new company commanders to "teach him how to do it." His focus was to train as we fight—therefore using performance standards, or measures of success, effectiveness, and performance that would become ingrained in all the army does. This shift in measuring training with definitive criteria was the most significant change he made in his tenure as the TRADOC commander, which he maintained later on as the Chief of Staff of the Army (CSA).

Taking its lead from TRADOC, the infantry school was tasked with developing new training doctrine with a focus on performance-oriented training. There is no doubt that KC Leuer had a great influence not only on the infantry school with his passion for POT, but also on the rest of the army. As stated, Leuer was disgusted with the way his army operated in Vietnam, believing there was clear potential for successful missions until they were translated into "search and destroy tasks." Leuer said that he can think of nothing more demeaning then to tell someone to search and destroy. Where and what? There was little or no definition in operational and tactical terms to provide a concise directive where people could do their jobs and leaders coldly assess the outcomes in some sensible way.

Leuer's intense disdain for unfocused mission sets with immeasurable ways of defining success led to a full-bore effort to implementing and honoring performance standards for everything the future Ranger force would do. As Leuer describes it, with POT we now can say good job, bad job, or "got to pick up" in certain areas—all with some level of intellectual assurance based on validated performance-oriented training and standards.

If the new Rangers were going to do what Abrams wanted done, everyone had to do a job they were supposed to be doing. This seems like a nonnegotiable fundamental for any work ethic; however, the army of 1974 had leaders at all levels not performing to standard. With

POT, squad leaders knew what they had to do, platoon sergeants roles were easily defined, and junior officers could finally be the junior leaders in charge of training.

Leaders like Leuer, at the time, realized training could be conducted with not just one person who may have been the only expert on the topic at the time who trained everyone on what *he* thinks is right as the expert on the winning standard in *his* mind. With POT infused in an outfit for everything it does, observable and measurable objectives are gained whereby any level of leader can train his own soldiers and small-unit formations to a universally understood mark. The newly energized leaders put in charge of the new Ranger formations were pioneers in this effort. POT was one of things that allowed for forward progress and gave the NCOs an opportunity to lead—a task they took freely and energetically and then never looked back.

Poor discipline, ambiguous mission sets, a disillusioned force, and unpopularity with the nation left our army in dire need of change. Something had to happen. The all-volunteer force was an important change to how our army was populated, but positive change because of the jettisoning of a conscript force, would take years if not generations to reap benefits.

The world was not getting to be a safer place. The Wall still existed in Eastern Europe, and the superpowers that existed at this time were on a collision course for conflict. The leadership at the highest levels in the military looked to the past in defining what kind of unit the U.S. Army turned to time and time again to form the force of choice during the most volatile eras in our military. Thus, in the fall of 1973 Gen. Creighton Abrams, chief of staff of the U.S. Army, issued a charter for the formation of the 1st Ranger Battalion.[25]

3

The Abrams Charter, the Ranger Creed and Manning the Force

Recognizing that I volunteered as a Ranger…
The first stanza of the creed defining the mindset of this
elite all volunteer force as forged by its founders.

SELECTING A COMMANDER

Lieutenant Colonel KC Leuer was assigned to Fort Benning after attending the Industrial War College in 1973. Soon after arriving at Benning he was sent to Florida State University's (FSU) Department of Educational Technology for eight weeks with the task to study performance oriented training (POT) by, in his terms "some of the best gurus in the nation."[1] This education was a fortuitous and timely pedagogical adventure for Leuer. It would serve him and the Army well in the very near future.

When Leuer returned to Fort Benning, following his education at Florida State University, Gen. (retired) William R. Richardson, the Assistant Commander for Fort Benning, gave him a very important mission based upon his new found education. Leuer said that Richardson directed him to train the instructors in a department headed by Colonel Billy Rutherford on the merits of performance oriented training (POT).

The Performance Oriented Training concept and its accompanying program of instruction (POI) was championed by the Department of the Army (DA) with a goal of transforming the US Army Infantry Center (USAIC) and other professional military education (PME) institutions to a task-conditions-standards (T-C-S) construct. DePuy with the support of General Paul Gorman had surrounded himself with deep thinkers from across the Army at this volatile time.[2] Not only was the US still involved in the waning years of the Vietnam War, It was during this time that the 1973 Arab-Israeli War was ongoing.

Strategists and operational thinkers were trying to make sense of this Middle East conflict that involved denial, deception, and the equalization of infantry in taking on tanks with anti-tank weapons. Addressing air superiority with saturation of surface-to air missiles (SAM) the war pitted western technology (Israeli) with Soviet hardware (Egypt and Syria). DePuy used the experiences, lessons learned and recognized threats to develop performance-oriented training – what we would today call "train as we fight."[3]

Influenced mostly by the lessons learned by this War and from the techniques of the Germans and Tactical Air Command, DePuy would eventually produce a manual focusing on high intensity conflict (HIC) in Europe and aimed to negate the bitter experience of Vietnam. By 1976, DePuy and his organization developed the new Field Manual (FM) 100-5, *Operations*. This manual attempted to integrate how the Army procured, trained, and fought under a completely new a concept called "Active Defense." DePuy's mentor General Hamilton Howze termed it as suppressive fires or "overwatch" as the best way to neutralize enemy fires so that maneuvers could take place in a decisive manner.[4] DePuy offered that Active Defense is more about intelligence gathering capacity to enable a force as opposed to an ability to gain and maintain a mobility differential like in a new type of track vehicle or helicopter.[5] This new type of operational thought had to take into account the change from a conscript to an all-volunteer force.

The last men the Army drafted were born in 1952. These men had to report for duty in June 1973, thus official ending the conscript force.[6] With this shift in how the Army acquired its soldiers came other changes to include a project to develop a Ranger Battalion whose focus was readying for combat using Performance Oriented Training (POT).

The action officer for this Department of the Army level project was Major William Spies. Spies said that Abrams directed that there was going to be a Ranger Battalion and not a Special Forces Battalion and not even just a Special Operations Battalion to generate an elite set of warriors in the post-Vietnam era. The reason, Spies said, that Abrams chose it to be a Ranger Battalion is that because every time he saw something outstanding happen "down there in Vietnam" was because a Ranger was doing it. Spies recalls that this probably was said sometime in the first week of December 1973.[7]

Spies was basically a bystander in the meeting that included Abrams, DePuy, Gen. Kerwin, the Commanding General or Forces Command (FORSCOM), Maj. Gen. Thomas M. Tarpley, the Commanding General of Fort Benning, and a Col. Hatch from the US Army Infantry Center's (USAIC) Combat Development Branch. However, right after the meeting he realized he would be asked to put together the Table of Organization and Equipment (TOE) for this new unit.[8]

To complete this daunting effort, Hatch solicited the support of Sergeant Major Jim "Snake" Collier, a former advisor to the Vietnamese 81st Airborne Ranger Battalion, the only Airborne Rangers in the RVN, and an experienced Ranger-qualified paratrooper. Collier would eventually become the Command Sergeant Major of the Ranger Department at Fort Benning and today is a member of the Ranger Hall of Fame.

Spies and his team helped the Infantry Center form the 1st Ranger Battalion. Not only did he develop the TOE, he also staffed the first Ranger Battalion operations field manual and as the infantry doctrine officer of the Combat Developments Command he integrated all things Ranger into the Army lexicon. He was a focal point in implementing changes to performance oriented training using his Vietnam experience with the 1st Force Reconnaissance Battalion to make training more realistic and valid. He also helped implement the change to task, conditions and standards (TCS) in the Ranger course.

This effort was quick. Once Abrams made the decision to form the battalion, Spies and his team had from late November 1973 until January 1, 1974 to develop the TOE and a draft set of missions sets that the Battalion would have to accomplish.[9]

Around Christmas time in 1973 KC Leuer first heard rumbles about development of a new Ranger unit. He first heard of this potentiality from Gen. Richardson's Executive Officer recalling that he simply said that "TRADOC is going to form a Ranger Battalion."

This development peaked Leuer's interest but being a naturally humble and unassuming man he hadn't thought of how it would personally impact his career until Richardson asked him if he was interested. Leuer immediately told Richardson that he would like to

Lieutenant Colonel
Kenneth C. Leuer, the first
commander of the modern
day Rangers.

be considered for the assignment. Two weeks later Leuer was told that Gen. Creighton Abrams, the Chief of Staff of the Army (CSA) personally "by-named" him as his choice for the command.

Ken Leuer was a 1956 graduate of the University of Iowa and a NCAA Wrestling Champion. He served in the 82nd Airborne Division, the 8th Infantry Division, the 2nd Infantry Division and the US Army Special Forces, before deploying to Vietnam with the 173rd Airborne Brigade for his first combat tour. Leuer then he went to the Pentagon briefly before heading back to Vietnam with the 101st Airborne Division for his second combat tour. With his professional breadth and depth of experience up to this point it is no surprise he was chosen to be the first leader to command a Ranger Battalion since World War II.

Leuer was a proven combat warrior and a tenacious leader. In 1972 he took command of 2-501 Airborne and immediately moved the battalion to Firebase Tomahawk, a vulnerable outpost south of Hue and Phu Bai. Tomahawk had been overrun the previous two Aprils and Leuer was determined to ensure it wouldn't happen again. This is when he first met Creighton Abrams.

Abrams had become the Commander of US Military Activity Vietnam, or COMUSMACV, replacing Gen. William Westmoreland. During a visit to Leuer's firebase, Abrams inspected its defensive posture. Upon inspection, Abrams corrected the placement of "tanglefoot" (the tripwire woven into a web close to the ground along the camps perimeter) directing Leuer to "tighten it up" and subsequently directing Leuer's commander to relieve him if the deficiency was not corrected in a timely manner. A confused Leuer learned later that tanglefoot and other base security measures had been Abrams pet peeves ever since an Americal Division battalion had been overrun the year before.

Leuer gives credit to his selection as the first Ranger commander to General Bernard Rogers, then the Deputy Chief of Staff for Personnel or DCSPER. Leuer was told that when Abrams asked Rogers who the first commander of 1/75 should be? Rogers said, "My number one choice is Lt. Col. Ken Leuer."[10]

After deciding he wanted the Ranger job it took two weeks before he would hear anything in return. Abrams saw the potential in Leuer and stood by his decision to select him. Leuer was told some time after his selection that the requirement to fill the position was staffed through Infantry Branch, the personnel command that controls the assignments of all infantry officers for the Army. The instructions that Infantry Branch received was that it directed to list five officers and stack their files one through five in order of most desirable for the command. Leuer's was listed as number one.

Once the files were selected and stacked in priority order, the DCSPER then brought the files to Rogers who then brought them to Abrams for the final selection. Abrams said he thought he knew Leuer personally and inquired if they had been in Vietnam together. Even though Abrams was the Commander of Military Assistant Command-Vietnam at the same time Leuer was a battalion commander in Vietnam and had experienced the "tanglefoot" visit to Leuer's camp it was never clear if Abrams had remembered him personally. At no time during the selection process did Abrams ever personally talk to Leuer about anything to include the "tanglefoot" incident.

In February 1974 KC Leuer was informed by Gen. Richardson that Abrams selected him and it was official. He was now the commander of the 1st Battalion, 75th Rangers (Infantry). The modern day Ranger era was born.

THE CHARTER

After accepting the command, Leuer's next big step was to man this prestigious organization and then determine how to proceed with making the Battalion meet the expectations of the senior leaders in the Army. Leuer would receive more guidance about the command in an already planned visit to Fort Benning and the Infantry Center by Gen. DePuy. DePuy had scheduled months prior to speak to all of the Army's school house commandants gathering at Fort Benning. This proved an opportune time for him to convey Abrams guidance to Leuer face to face.

Leuer was told in the morning of the visit that DePuy wanted to see him at Lawson Airfield. He waited all day for the meeting and finally met DePuy in late afternoon around 4 pm. Leuer was nervous and paced back and forth waiting for DePuy to arrive. According to Leuer the two men met in private and DePuy asked him "well, what do you think is expected of you?" Leuer's response was that he heard the battalion was supposed to be the best-trained infantry in the world, and that's all he heard. But he did go on to state that he believed that they were supposed to set the example of what the Army should be as it pulled out of the doldrums of Vietnam. DePuy went on to tell him that the US Army is in the worst shape it's been in a long time.

Lt. Gen.(Retired) Gary D. Speer, a Platoon Leader and Company Executive Officer for Company A and the logistics officer or S-4 for the 1st Battalion (Ranger), 75th Infantry from 1974-1977 agrees with DePuy's characterization of the Army in the 70s and places things in context about the Charter. To Speer, the US Army coming out of Vietnam represented many challenges, but he is very careful about demeaning the great soldiers who continued to build a better Army during that difficult time. Speaking of Abrams, Speer thought he was truly the right leader at the right time for our Army and our Nation to lead the Army out of Vietnam and set the course for the future. Although an armored officer by background, Abrams' vision for the future placed much weight on the shoulders of light infantry in the decision to form the 1st Ranger Battalion

After DePuy told Leuer that was selected he conveyed to him the CSA's guidance. This "guidance" became widely known as Abrams Charter. However, the so-called Charter didn't come directly from Abrams to Leuer and was given only verbally to Leuer by DePuy. Leuer listened and wrote the guidance on his 3 × 5 cards and later rewrote it the way he heard it said to him by DePuy. Leuer admits that "there is no such thing as a written Abrams Charter. There was no tape recorder, nobody listening to take notes for him, just the cards from which I wrote what I heard." Paraphrasing to an extent the guidance that came from DePuy as conveyed by Abrams the following is how Leuer understood the Charter:

> The Ranger battalion is to be an elite, light, and most proficient infantry battalion in the world; A battalion that can do things with its hands and weapons better than anyone. The battalion will not contain any "Hoodlums" or "Brigands" and if the battalion is formed of such persons, it will be disbanded. Wherever the battalion goes, it will be apparent that it is the best.[11]

In short, the Ranger Battalion was to be the "Gold Medal Infantry of the world."[12] And much like the efforts of other Ranger units from other eras, the members of the Ranger organization were chartered to go throughout the Army set the example for professionalism in how to train, how to live and how to fight. Other guidance Leuer took from that meeting was that if, in any way, the Battalion disgraces the United States Army or the US as a nation, the commander and the commander alone will bear the consequences. Finally, DePuy told him that Leuer needed to develop a Creed that the Rangers will live by, train by, and fight by.

DePuy told the new commander to keep his boss informed and he will do his best to keep Leuer informed, and that if he had problems he couldn't solve and he didn't tell DePuy about them then it's his fault. Leuer left that meeting fully understanding that he was in charge, responsible and accountable.

Distinctive 1st Ranger Battalion Scroll.

He also said that the establishment of this new unit was going to happen and you are going to be held accountable to make it happen. Both DePuy and Abrams knew that the establishment of this new elite unit took a lot of effort from throughout the Army to include experts and proponents in training, operations and leadership. Leuer felt that he had the Army available to support of this effort. DePuy told him he had an open check book and free access to recruit anyone in the Army. The responsibility was now placed on Leuer to make it happen.[15]

ESTABLISHING THE FORCE

The 1st Battalion (Ranger), 75th Infantry was activated on January 25, 1974 with an effective date of January 31, 1974. Abrams vision was that this Battalion would be molded after the 5th Battalion of World War II fame.

General Oder 127 was the directive to establish the Battalion. The recruitment and selection of soldiers Army-wide began immediately. On March 5, 1974, the Commander Military Personnel Center released a message to the field recruiting solders for this new Battalion. "The message read in part:

1. The Chief of Staff, Army has directed the establishment of a Ranger Battalion designed to be the finest foot infantry battalion in the world. The battalion must be capable of accomplishing any infantry platoon, company, or battalion type mission to include participation in airborne, airmobile or amphibious operations. The unit has been designated the 1st Battalion (Ranger), 75th Infantry.
2. Enlisted personnel who desire assignment to the 1st Battalion (Ranger), 75th Infantry, are requested to submit volunteer applications indicated herein.
3. Those personnel submitting a volunteer application must meet the following criteria:
    a) Volunteer for airborne and ranger training if not already so qualified.
    b) Regardless of prior airborne/ranger qualifications, be able to meet the medical, physical, and mental prerequisites for attendance at the Airborne and Ranger Schools …
    c) Have no record of GCM [General Courts Martial], not any courts-martial during current enlistment."[13]

Abrams desperately wanted this Battalion to be a leader of change for the rest of the Army. Ken Keen (former Regimental Commander when the author served with the Regiment as the Regimental Executive Officer and now a retired Lt. Gen.) did a study titled "75th Ranger Regiment: Strategic Force for the 21st Century" where he researched the benefits derived from this change transfer vision. He writes, "It was understood that the Rangers were to be a role model for the Army and leaders trained in the Ranger Battalions should return to the conventional Army to pass on their experience and expertise."[14]

Initially, Fort Stewart, Georgia, was considered the battalion's home but the Battalion and its cadre would undergo an assemblage of equipment and men at Fort Benning, Georgia. The first wave of personnel equipment were to come from Company A, 75th Infantry of the 1st Cavalry.[15] However, by all accounts that plan never came to fruition.

In the end, Ron Rokosz recalls that the majority of, or at least the initial cadre, came from the already formed Ranger Department at Fort Benning and the 101st Airborne Division because of Leuer's ties to that Division in his previous combat tours. The remainder of the initial personnel feeds came from a smattering of other units across the Army. The equipment, in the end, came to the 1st Ranger Battalion from Fort Benning stocks and as the Department of the Army saw fit.

With the decision to form this great battalion being born with its own Charter and expectations for success, it was now time to man the ranks. Considering that the 1st Ranger Battalion was chartered to be the best example of what we want in the Army, the men would serve in it had to be the best of the best. By assembling the best possible talent to serve as Rangers, these men could infuse their professionalism and high standards throughout the rest of the Army after serving their tour with the 1st Battalion.

Setting the bar high for the rest of the army required a talent pool unmatched and unequaled. To be the model fighting force Leuer knew they had to make the mark through leadership and dedication and good people.

As the new commander of the 1st Ranger Battalion stationed temporarily at Fort Benning, Georgia Leuer reported to Richardson and maintained a TRADOC command relationship through the Army Training Test or ATT in December 1974. Once it was trained and available to deploy the 1st Battalion would fall under the command of US Forces Command (FORSCOM).

Leuer got a lot of guidance in the beginning from everyone to include a lot of unsolicited advice and guidance on everything from how much you should carry in a rucksack to how many socks you should carry on a road march. But the best guidance he got was from Richardson whom Leuer had held in high regard.

Lt. Col. Keith Nightingale, Plank holder of the 1st Ranger Battalion.
(Source: Nightingale)

Very early on in the command Richardson told him he will get the option to acquire the best people and was told he would be offered people all the time as well. Everyone is going to have a "favorite son" who they would try to push on Leuer conveying he "is the greatest thing going." But Richardson told him that YOU have to be comfortable with whom you select for they have to be loyal to you and if you don't feel comfortable with someone then don't take him. Richardson told him to take those that you think sees the same thing you do and take those that you would want around in peacetime and in a fight. Leuer thought that this was the best guidance anyone gave and appreciated the timeliness for he received it just prior to visiting Personnel Command (PERSCOM) on how the selection process would go.

It seemed to be an easy prospect; Leuer would select who he wanted and the owning unit would give them up. After all, this was the US Army Rangers and everyone wanted to serve as a Ranger or at least support this elite fighting force. Leuer would find out very soon that that wasn't the case. In fact, the Rangers were not well respected or at least many were leery of their prospects to form a unit characteristic of what Abrams expected.

Col. (retired) Jerry Barnhill, the original C Company Commander was notified of his selection by Infantry Branch. He had been serving in the Department of Airborne Tactics at Fort Benning when he received the call. Having had two tours in Vietnam with the 101st Airborne Division, many thought, to include many branch officers that by accepting this assignment he would be wasting his time. Barnhill accepted the offer and never regretted it.

Brig. Gen. (retired) Ron Rokosz recalls that he was also stationed at Fort Benning and assigned to the Department of Tactics when he first heard of the new Ranger unit. Rokosz was in a firing tower at a rifle range reading a current copy of *Army Times* where a headline read that the US Army was forming a new Ranger Battalion. The article said that all that joined the team would be hand selected, on parachutist status, fitted with camouflage fatigues and receive specialized training.

He vividly recalls being excited about the potentiality of becoming a Ranger and remembered saying to the officer in the tower with him: "Wow, how would you like to get selected for that unit?!" Ironically, once he returned to his main office later that day he was told that Infantry Branch had just called. He immediately agreed to interview with Leuer who was also at Benning at the time and soon after became the first B Company Commander for the Battalion.

Col. (retired) Keith Nightingale was the original commander of Headquarters and Headquarters Company, 1st Ranger Battalion and the eventual Ranger Training Brigade Commander. He served two tours in Vietnam as a company commander for the 101st Airborne Division and a senior advisor to the 52nd Battalion, Vietnamese Rangers. Nightingale knew Leuer from the combat zone in Southeast Asia and the 101st placed a call to the new Battalion Commander requesting Nightingale's assignment and was selected immediately.

After arriving at the Battalion someone asked Nightingale if he knew Rokosz, but he did not. However, when they first met they took an immediate liking for each other and served as brother company commanders during the initial standup of the Battalion. Nightingale did know Col. (retired) Jim Montano who would become the plans officer or S5 before eventually taking command of C Company, 1st Rangers. That is how it worked, those that knew the boss had an "in" and those that knew each other had the fortune of having the boss swayed to giving them a shot.

To Richardson, Fort Benning was the right place for the initial formation and cadre training. Fort Benning is rife with resources for it is a major US Army training base. A post like Fort Benning had no problems providing training resources, school slots, and other support. Richardson left the details with all of the resource management and synchronization of those resources with then-Col. David Grange (who was the commander of the Rakkasans or 3rd Brigade, 101st Airborne Brigade in Vietnam, the same brigade the author commanded some 35 years later) who was Director of the Ranger Department at the time.

FINDING THE RIGHT PEOPLE

Recruiting was an adventure marked by many challenges and a vital number of successes. When recruiting lieutenants, the junior officers were asked who they knew and who their commander was and then a phone call was placed. At first the process was very fast. As Leuer puts it, the "machine ran on" and the initial fill of cadre was very quick. There was a major push to get approximately 600 Rangers into the ranks as soon as Leuer could. He had hoped to get the vast majority on board prior to the initial activation-type jump into Fort Stewart, Georgia which was planned for later in the summer 1974, but up until that day they were still acquiring people.

After being hired as the very first Ranger Battalion commander Leuer had set two simple priorities. These were: 1. getting good people hired and, 2. implementing performance oriented training (POT).

To meet the first priority Leuer went all out. Like any good commander it was important for Leuer to enlist the help of his senior noncommissioned officer in devising a plan for acquiring the right talent pool. That senior noncommissioned officer for Leuer was Command Sergeant Major (retired and now deceased) Neal R. Gentry.

According to Leuer, CSM Gentry was an enigma and a story in and of itself. The Command Sergeant Major of the Infantry School at the time, CSM "Bull" Gergen was a First Sergeant in 3rd Battalion, 506th Infantry that went to Vietnam with 3rd Battalion, 503rd Infantry when Leuer was the executive officer for the battalion. Leuer met him on the troop ship on the way to the Republic of Vietnam (RVN). He was asked by Leuer to be the Command Sergeant Major of the Rangers but he opted to stay with the Infantry Center assignment.

This was a time when Command Sergeant Major status was just becoming bestowed as command related and Neal Gentry was not on the Command Sergeants Major list. Regardless Gergen believed Gentry was the right man for the job and offered his name to Leuer. Leuer found out that Gentry was physically broken in many ways and couldn't run and "do a number of different things" but was the right guy for a number of reasons. In the end, Gentry was able to instill a focus in recruiting the right soldiers and noncommissioned officers and imparting discipline in the ranks.

Overall, Gentry was unorthodox but highly respected. All the company commanders knew that the key to success for team building was finding the right command sergeant major. Nightingale remembers thinking that "… many of us were somewhat taken aback initially when it was clear he wasn't in good shape and had timely 'absences,' but over time, we grew to like and respect him as a person. He did all the road marches and field training exercises and was, I believe, crucial in getting the Ranger noncommissioned officer model jump started."

In his initial push to attain the "right" people, Leuer created recruiting teams to visit each FORSCOM and TRADOC post in the Army. He estimates that his teams visited ten different places in the continental US in the first few months. The team usually consisted of a couple of the first sergeants (1SG) and a company commander. CSM (retired) Don Bruce was a platoon sergeant in B Company, 1st Ranger Battalion in 1974. He recalls going on recruiting trips to Fort Lewis, Washington with then-Capt. Rokosz and 1SG Jack Schmidt.

Shortly after arriving to the Battalion, Ron Rokosz headed a recruiting team consisting of himself and two noncommissioned officers to Fort Lewis, Washington. Like most highly advertised recruiting trips Rokosz gave a general briefing on the battalion, then had one-on-one interviews with all interested junior enlisted to review their personnel files. Anyone with disciplinary issues in their record was not accepted. That was the rule.

As stated earlier, the US society was rife with racist overtures as African American attempted to find their way as equals in a predominantly white society. The ongoing rights movement honored the respectful service many African Americans gave to their nation, even if it meant serving in the unpopular war in Southeast Asia. The military habitually mirrored American society and the Army was a microcosm of the crises and challenges Americans were experiencing at the time. And one of those challenges was indeed racism.

Leuer recognized a need for equality amongst his ranks and acknowledged that African Americans were serving proudly throughout the Army. Therefore he worked hard to hire at least one company commander and first sergeant who were African American. As a result, the Headquarters and headquarters First Sergeant, Willie Cobb (retired as a Command

First Sergeant Willie Cobb. (Source: Currie)

Sergeant Major and now deceased), was African-American as was Capt. Don Clark, the first original company commander for Company A. Additionally, the headquarters signal noncommissioned officer was an African American as were some of the cooks. According to Nightingale, the effort to draw upon a diverse population was "an interesting" and challenging issue. When the recruiting teams went on recruiting trips it was always a desired outcome to provide opportunity for all to serve equally in the Rangers, but the standards were never diminished to meet a quota.

MSG (retired) Don Feeney recalls his first interaction with 1SG Cobb. He had just arrived to Fort Stewart. As he entered the doorway of building 18, there stood 1SG Willy Cobb, black as coal and big as hell. The next few minutes changed his life forever. 1SG Cobb was wearing highly starched "cammies," pistol belt and a black beret. The first words out of his mouth were, "SGT Feeney, so you want to be part of the First Ranger Battalion started since World War II? It will be the most elite infantry fighting unit in the world. You think you're good enough?" Stunned and surprised he said "yes", not really sure what he was getting into. But he figured that 1SG Cobb sure looked fit to fight. He was very impressed with his whole demeanor and wanted to be just like him.

1SG Cobb said "Well SGT you just volunteered for the 1st Ranger Battalion. You're to be at Fort Stewart, Georgia by next Thursday so go home and pack your shit." He asked him where Fort Stewart was and where was the Ranger Battalion located on it? He told him, "You're a Ranger, go find it."

He would later serve under 1SG Cobb in Charlie Company, 1st Battalion, 75th Rangers and never forgot his first meeting. To this day Feeney is still trying to live up to the standards of leadership and honor that Cobb and other leaders like him set in the 1st Ranger Battalion.

In reality, 1SG Cobb conveyed to Nightingale that it was very difficult to recruit African American soldiers at times. There was no doubt that there was a huge population of courageous and talented soldiers of all race and ethnic background but one thing the Rangers would not do is violate their standards. One of the standards was an ability to pass the swim test, a rigorous examination of the soldier's ability to swim both freestyle and in combat gear. Most of the African Americans serving in the military during this time were from the inner city where swimming just wasn't something that was taught to city children as they grew up. Therefore many good African American soldiers were rejected for they just could not swim. The Rangers were not going to waiver on their standards. The standards were the standards and nothing, no matter how sensitive the desire may be was going to violate the Ranger standard.

Besides briefing the potential recruits on the concept of what it was to be a Ranger in a Ranger unit, the team was first required to arrive on Post and meet with either the Chief of Staff or personnel officer (G1) and brief him on the plan. A typical visit lasted a week in an effort to attract all grades.

Making the Ranger candidates understand that the Battalion was serious about making the noncommissioned officer the heart of the program was critical. Having gone through the now-organized Ranger Assessment and Selection Program (RASP) board as a venue in today's Ranger Regiment for interviewing new Ranger Candidates, I can say with all certainty that that reliance on the noncommissioned officer as the backbone of the Ranger Organization is alive and as well as it has ever been.

Leuer recalls that some of the major shortfalls early on were Staff Sergeants (E6s) and Sergeant First Classes (E7s) as well as junior enlisted men. According to Keith Nightingale, finding officers, senior noncommissioned officers and junior enlisted was pretty easy. Finding midgrade noncommissioned officers was very hard for every command wanted to keep their finest and most promising NCOs.

The other obstacle was the stubborn and oftentimes covert opposition of the "Institutional General Officers" commanding the posts where they recruited. Naturally, they didn't want to allow their best troops to leave to join the Rangers and covertly frustrated recruiting in a variety of ways. Even though the Chief of Staff of the Army (CSA) directed that the Ranger Battalion would be "fully supported and encouraged" by posts Commanding Generals and their subordinate commanders and staff, Leuer would run into some major roadblocks and friction in his efforts to man his force.[16]

Typically the briefing made it very clear with a discussion of Ranger standards, the Physical Training (PT), road march, performance oriented training (POT) and over all disciplinary requirements that a potential Ranger assignment would be challenging. They also made it very clear of the importance of the noncommissioned officer and the reliance on them to be the standard bearers. The bottomline message was that it was a tough, demanding assignment but if they made it that could be part of the finest light infantry battalion in the world.

The plan for drawing in new Rangers was good, but its execution had issues. While the CSA message required the visited post to provide those interested soldiers without interference there was much push back from a number of commands to support the program.

The friction the recruiters felt was none more apparent than with Leuer's former Division, the Division he went to war with and commanded in so courageously in combat; the 101st Airborne Division. Leuer explains when comparing how receptive certain commands were that the only one that he really "got nailed by" was the Commanding General of the 101st – Sid Berry. The reason why he sent recruiters there first once he was notified how the process for recruiting would work is because Berry was the assistant division commander (ADC) of the 101st in Vietnam where he had known Leuer for six months in combat.

Berry would visit Leuer in Vietnam telling him that that he felt he was doing things well, so naturally Leuer thought he had a real friend in Berry and felt a level of comfort going to Fort Campbell to recruit from the 101st's pool of would-be Rangers. However during the Rangers courtesy call he told Nightingale, who was representing the Rangers, that all the great soldiers are in the 101st and "you aren't get any of them." This poor reception, or as Leuer describes it, "when Berry fired that salvo on me," left Leuer reeling realizing this quest for good recruits was going to be a tough road. To add to the angst was that Nightingale had also just come from the 101st and was on tap to be Berry's aide when he opted to go to the Rangers instead.

Interestingly enough, Berry was one of the few Ranger qualified generals at the time and Leuer figured of all commanding generals he would jump at the chance to support this new Ranger formation. Apparently, Berry hadn't read what Abrams directed or if he did, it wasn't distributed or even more likely it wasn't believed.

The post leadership, wherever they went, usually worked hard behind the scenes to discourage access to Ranger recruiters and to minimize the availability of troops to attend. It got so extreme at times that some of those recruited were threatened that it would have an impact on their efforts and jeopardizing their careers by disrupting the institution. Commanders, in turn, would threaten their subordinate commanders by making it clear that the commander would look at any loss of any soldiers as a reflection on that unit's leadership.

In reaction, Leuer immediately took a quick trip to the nation's capital and told the "powers to be" at the Deputy Chief of Staff for Personnel's Office (DCSPER) about the lack of support he was getting from the field at forming the new unit. Leuer didn't pull any punches in the initial feedback for this mission was too important to fail so he explained in detail what had occurred at Fort Campbell and immediately the "world changed overnight." At Fort Campbell, Kentucky the recruiting effort improved dramatically. At first recruiters were not given a lot of support. There was no success on the first two days. After Leuer went to Washington, DC the third day realized great success with 350 people lining up for interviews. This also when the nomination followed by an interview process began.

The arrangement Leuer made with the Military Personnel Center or MILPERCEN was that they would nominate a potential Ranger officer or noncommissioned officer and Leuer would interview them and then make a decision. It was Leuer's authority to accept them or not. The files were sent to Leuer; he would review them and decide to see the potential Ranger. This arrangement worked with MILPERCEN for the better part of 1974. Those that weren't accepted were merely sent back to their units with no repercussions; must like it's done today.

As the selection process went along there were very few soldiers being picked from the Army's 82nd Airborne Division, a unit that was rife with airborne qualified soldiers. Being airborne qualified by completing a three week intensive school where future paratroopers made five mandatory jumps to attain their coveted silver wings upon their chest, was a requirement to serving in the Ranger Battalion. However once the announcement of the new Ranger Battalion was made there was apprehension at Fort Bragg and the 82nd held their people. Between the 82nd Airborne and the US Marines there was great concern and posturing for both elements thought they were the US "ready force." They both felt threatened.

Like at the 101st initially, the 82nd didn't want to lose its people. While the early stages of recruiting proved difficult for the 1st Battalion to acquire 82nd Paratroopers, the 82nd would eventually concede. However by the time they "saw the light" it was too late for the 1st Battalion and most of the Bragg soldiers went to the 2nd Ranger Battalion which would be activated on October 1, 1974 at Fort Lewis, Washington.

Commanders kept telling Leuer that they didn't want to lose their talented people. To Leuer this was ridiculous for the new Ranger Battalion only needed a few hundred talented soldiers and if that was the only amount of "talented people" in a 20,000 man division then their problems were greater than losing a few hundred to the Rangers. Leuer could understand the issues with taking soldiers from a command, especially the good ones. However what he couldn't understand was that the Army had 1.4 million men at the time. He needed 600 and he was hearing stories and people were telling him he was stealing the best from each unit. What his confusion was that if he took the best 600 people out of the 1.4 million clearly the remaining 1,399, 400 had some talent.

Leuer looks back now and regrets not personally calling each commanding general of each post in his effort to find the right soldiers. However, Leuer was "just" a lieutenant colonel and it was difficult for a commander, four ranks removed, to personally call a two-star general asking him to talk to and potentially grab his people.

In the end, the young soldiers Leuer and Gentry got to come to the Battalion were full of energy and excited about being part of the elite. But not all Rangers were six foot tall and chiseled. It took all kinds, as it does today to make a warrior. Some of the most unlikely men turn out to be the steely eyed killers on the battlefield. Training and standards are what make a Ranger unit.

SGM (retired) Don Bruce was one of those unlikely heroes. The first noncommissioned officer he met upon arriving at the Rangers was SSG George Conrad. They became fast friends and Ranger Buddies. On the third evening after his arrival he and Conrad went out for a beer. Conrad knew the lay of the land at Fort Benning and its sister city, Columbus, and told Bruce there's only one place really for us; it was called Donatos. This establishment is where all the Ranger committee members, Black Hats (Airborne Instructors), Pathfinder department instructors and other top of the line soldiers hung out.

Bruce and Conrad went down a dirt road and stopped at the little shack hosting Donatos. They walked in and several of these seasoned warriors greeted Ranger Conrad. Right away George introduced Bruce around.

Donatos had a dirt floor except for the plywood that made up the floor to support the pool table. They had two heavyset barmaids and only served long neck Budweiser. The bar stools were commodes and the bar was made of plywood. Bruce went to the pool table and soon several other Rangers from 1st Battalion arrived.

Bruce had just broken the rack when the bar maid came to deliver our beers. Bruce looked up to say thank you, and her eyes opened wide. She dropped her tray of long necks, grabbed the cheeks of his face and shouted "LORD HAVE MERCY, YOU'RE OPIE FROM MAYBERRY." The whole place roared with laughter as Bruce tried to explain that no, he wasn't. Even though he was 24 at the time he looked like Ron Howard's twin.

Bruce had that young baby looking face and he and Howard's haircuts where the same. In the past, people had commented about the similarity in looks between them, but nothing like her. And with his new-found Ranger buddies it was not the right setting for a bar maid to make such an uproar.

Bruce, however, simply helped her pick up all the longnecks but the rest of my new Ranger buddies couldn't stop laughing. The next day the story went like wildfire through the noncommissioned officer ranks of what happened the night before. And Bruce's nickname was born. Forever he would be known as Opie; but he was still a US Army Ranger.

An example of where the recruitment was an unqualified success can be found in excerpt of a letter sent to General Abrams by Private Edward P. Lariviera. At the time, March 23, 1974, Lariviera and his buddy Private Danny J. Minnick were assigned to the 1st Battalion, 505th Parachute Infantry Regiment at Fort Bragg, North Carolina. Lariviera writes:

> Sir. I am writing this letter on behalf of myself and Private Denny J. Minnick in regards to the new Airborne Ranger Battalion that is being formed in Fort Stewart, Georgia. When Private Minnick and I enlisted, we both wanted to be Rangers… when we were informed of the need for volunteers for the Ranger Battalion we were both very excited. We thought this would be our long awaited chance to be, or at least try to be, Airborne Rangers…

Since they were both too junior to apply they were asking the CSA for a rank waiver which they received. This letter meant so much to Abrams that he kept it and it filed with his personal papers collection.[17]

Leuer wouldn't have traded any successes he had in hiring those that were key to the Battalion's smooth activation and train up. He and his leaders clearly were getting quality officers, noncommissioned officers and soldiers. But before the cadre training and unit preparations for validating the unit's mission readiness could be executed Leuer had one important task the placed on himself and the Rangers; development and internalizing The Ranger Creed.

THE RANGER CREED

Maj. (retired) Todd Currie, an integral supporter of this book, and along with Ron Rokosz and Keith Nightingale, were a major part of the genesis of the idea for writing it, enlisted in 1974. He was immediately recruited by a Ranger noncommissioned officer who visited the basic course he was attending. At a very young age Currie saw a black and white movie about the Rangers at Anzio in World War II and, as a result, felt he always wanted to be a Ranger. He was born to a patriotic family that had a long history of serving in the military.

Currie thought that the Ranger Creed was more than words – its words included the concepts, ideas, and rules and basically was the crux of what it meant to be a Ranger. The way it was integrated into the Battalion, by Leuer and his senior leaders, left everyone aspiring to live by it. Officers and noncommissioned officers internalized it and it became a part of everything they did. This included a daily reinforcing ritual of reciting it to ensure its impact was felt by the entire organization.

According to CSM (retired) Mike Martin, the original B Company First Sergeant (1SG), everyone had to know the lineage, heritage and Ranger Creed. Even promotion boards had questions on the history, lineage and Creed imbedded in them. But it never seemed to be a problem for it was tested each morning at the initial formation. There is no doubt this simple

Ranger Todd Currie.
(Source: Currie)

six stanza creed was critical to the formation from day one. Its development rested on the shoulders of the most senior enlisted man in the Battalion, CSM Neal R. Gentry.

The Creed was tethered to Abrams Charter. The original Ranger Creed was developed by Gentry in 1974, but variations of the stanzas have occurred since. These variations are due to the diverging histories of the Rangers as a unit and Ranger School and how it evolved. After the consolidation of the Ranger Companies, the phrase, "of the Rangers" was substituted for "of my Ranger Battalion" by members of 1st Ranger Battalion. When 2nd Ranger Battalion was formed, they did likewise. After the formation of the 75th Ranger Regiment in 1984, members of all battalions adopted the wording, "of my Ranger Regiment".

Based on the history of the Ranger as described up until this point in the book, the establishment of the Charter and the Creed are important points in the path to success for the Rangers in the years since. The Ranger Creed is a six stanza document where the collection of each first letter for each stanza spells out the word RANGER.

"Recognizing that I volunteered as a Ranger, fully knowing the hazards of my chosen profession, I will always endeavor to uphold the prestige, honor, and high esprit de corps of my Ranger Regiment.

Acknowledging the fact that a Ranger is a more elite soldier who arrives at the cutting edge of battle by land, sea, or air, I accept the fact that as a Ranger my country expects me to move further, faster and fight harder than any other soldier.

Never shall I fail my comrades. I will always keep myself mentally alert, physically strong and morally straight and I will shoulder more than my share of the task whatever it may be, one-hundred-percent and then some.

Gallantly will I show the world that I am a specially selected and well-trained soldier. My courtesy to superior officers, neatness of dress and care of equipment shall set the example for others to follow.

Energetically will I meet the enemies of my country. I shall defeat them on the field of battle for I am better trained and will fight with all my might. Surrender is not a Ranger word. I will never leave a fallen comrade to fall into the hands of the enemy and under no circumstances will I ever embarrass my country.

Readily will I display the intestinal fortitude required to fight on to the Ranger objective and complete the mission though I be the lone survivor."

—Ranger Handbook SH 21-76[18]

As Keith Nightingale so eloquently puts it, "throughout history every nation has had a small military organization that is so utterly reliable and totally dependable to accomplish the mission of the moment or disintegrate itself trying… The Greek Hoplites, Caesar's Tenth Legion, Napoleon's "Old Reliables," the British Household Guards and the American Ranger – each had a Creed as the basis for binding and bonding their membership into a cohesive mass that thought and acted alike, shared common values and forsook privations while relegating personal desires for the greater good. In the darkest moments and in the most tenuous of times, the Creed focused the members of these formations and caused ordinary people to achieve extraordinary things. It was the foundation of the organizational soul."[19]

Initially, CSM Gentry got the mission from Lt. Col. Leuer to write a specific Creed for the Rangers. Leuer told him, "Sergeant Major, one of the items on the Charter I got from

General DePuy is we got to have a Creed. I want you to put together the Ranger Creed. I will edit it, I will approve it, and I will pass it on. You put it together."[20]

In Leuer's mind's eye it was intended to be the code by which each Ranger lived. He wanted to put something on the backs of the noncommissioned officer that could get them to "buy in" to what it was they were doing and the importance of the Battalion. Leuer describes this intention for he wasn't sure with what the Army had brought out of Vietnam that he could get the stimulus going that people wanted to be in – what he calls – the Gold Medal Ring. Nightingale describes the motivation for developing the Creed as Leuer's desire to have a single encompassing Creed that could put everything together for all ranks and be the foundation for how the organization operated and lived. Leuer gave Gentry the guidance to review other military adages and documents like the Code of Conduct to see if he could develop a code that encompassed all that is Ranger and to see if the word "Ranger" could be embodied in this code.

Gen. (retired) William "Buck" Kernan who retired in his last assignment as the Commander of the Command of the Joint Forces Command, is most known in the Ranger Community as the commander of the Ranger Regiment during Operation Just Cause in Panama in 1989. He was also a field grade officer in the Rangers in the late 1970s. He recalls that at first he did not have an appreciation for how much the words in the verses would mean in his professional development. He admits that he was more concerned about committing it to memory for fear of professional embarrassment should he fail to recite a stanza when called upon. He goes on to say how he would dwell on the meaning of the words and routinely found them inspiring during difficult times. In fact, during the combat jump into Panama, Kernan had each aircraft recite the Ranger Creed while the airframes were on short final three minutes out from the jump.

But just knowing the Creed by rote memory like Kernan alluded to was not good enough. Ron Rokosz believed from the start that he wanted it to be more than something one could quote, instead it should be more something Rangers could live by and feel bonded to.

Gentry came up with a draft based on Leuer's intent. Nightingale says that Leuer thought it was a good draft but needed work. The original Creed was heavily airborne-oriented and, according to Nightingale was a pretty close version of the Airborne Creed found posted up all over Fort Benning at the time. Leuer took issues with its initial length and the fact that it wasn't specifically focused on just the Ranger community.

The Battalion Executive Officer, Major Donald "Rock" Hudson took what Gentry wrote and began its transformation to a Ranger focused Creed. The finalized version for that period was reviewed at a Commander's Call (a get together of all commanders and senior staff in a unit) and then was distributed once the Battalion arrived at Fort Stewart, Georgia; it's new home.

This was the original draft that the Rangers originally implemented. The differences from the Creed shown earlier, the one that exists today in *The Ranger Handbook*, are noted in *italics*. Those words that didn't exist in the original draft but found their way into *The Ranger Handbook* are denoted by brackets [ ]:

> Recognizing that I volunteered as a Ranger, fully knowing the hazards of my chosen *service, and by my actions* [I will] always [endeavor to] uphold the prestige, honor, and high esprit de corps of *my* [the] Ranger Regiment.

Acknowledging the fact that a Ranger is *not merely* a more elite soldier who arrives at the cutting edge of battle by land, sea, or air, [I accept the fact that as a Ranger my] *but one whose* country expects [me] *him* to move further, faster and fight harder than any other soldier.

Never shall I fail my comrades *by shirking any duty or training, but* [I will] always keep myself mentally alert, physically strong and morally straight and [I] will shoulder more than my share of the task whatever it may be, one-hundred-percent and then some.

Graciously [Gallantly] will I show the world *by my courtesy to superior officers and noncommissioned officers, by neatness of dress, by care of my own equipment* that I am a specially selected and well-trained *Ranger* [soldier. My courtesy to superior officers, neatness of dress and care of equipment shall set the example for others to follow.]

Every day I shall respect the abilities of my enemies, but remember I am better trained and can [Energetically will I meet the enemies of my country. I shall] defeat them on the field*s* of battle for I *shall* [am better trained and will] fight with all my might. Surrender is not a Ranger word *and* I will never leave a fallen comrade *whereby he may* [to] fall into the *enemy* hands [of the enemy] and under no circumstances will I ever embarrass my country.

Realistically [Readily will] I *shall* display the intestinal fortitude required to fight on to the Ranger objective and complete the mission though I be the lone survivor.[21]

It is obvious of the changes that have occurred over the years. The lead words for some of the stanzas have changed and there is more of an emphasis on the personal embracement in the wording of the newer Creed so that the individual Ranger embraces it by saying more "I's." Thus the newer version denotes ownership by each Ranger. However, in 1974 the Creed was the Creed and it worked for Leuer and his command. It was embraced by the Rangers and they heeded every stanza.

Every soldier in the battalion had to know it by heart, but when it was first disseminated no one was informed that it had to be memorized. The importance of this Creed, however took hold almost immediately. The Battalion had physical training (PT) formation every Monday morning and for the first several weeks Leuer would stand on the PT platform and unrehearsed and even unplanned he figured he'd get his outfit to recite the Creed he so judiciously developed.

"Repeat after me, the first stanza of the Ranger Creed!" he exclaimed and then recite it in unison with the complete field of Rangers. They harmoniously would recite it with him. After several weeks of reciting the entire Creed himself in concert with his Rangers he began to call on other leaders to lead in a particular stanza. Initially many failed and the professional embarrassment – as hinted by Kernan – caused them to regroup and be better prepared for the next calling.

Eventually Leuer could stand on the PT platform and say: C Company, third rank, fourth soldier back, lead us on the third stanza of the Ranger Creed and that soldier would do so. For everyone in that formation it became a matter of pride and Ranger confidence to be perfect when it came to the Creed.

The Ranger Creed became a way of life. When he first had the vision for developing such a bonding mechanism he had no idea the depths it would weave itself into the formation and how much of a lasting effect it would have on the Ranger formations that would follow.

Today it can be heard before each mission a Ranger embarks and as the last formal organized task at the end of each operation.

CSM (retired) Mike Wagers who was a platoon sergeant in 1974 sums it up when speaking of the lasting effect the Ranger Creed had on the rest of the Army. He said that in the years after his time with the 1st Ranger Battalion he saw the change throughout the Army and witnessed the positive effect it had on a platoon and a company. If you were going to do it, do it right or don't do it at all. If it was wrong, then fix it and then do it right. Too often in the army he saw the easy way taken, the lesser hill climbed and the shorter route taken and then watched as commanders and troop leaders alike patted themselves on the back for accomplishing a task or mission without meeting the standard.

That didn't happen in the 1st Ranger Battalion and he did his best to insure that he never allowed it to happen throughout his career in other units throughout the Army. He passionately summed it up stating he was never as good as I was in those days, at that time, with those soldiers behind him and his leaders and commanders in front of him, and his friends beside me. "I was never that good again."

In 2001 in Misirah, Oman the author partook in such a ritual. Prior to the 75th Ranger Regiment, commanded by then-Col. Joe Votel (now a General and in command of the US Special Operations Command) loaded their MC130 Aircraft bound for the now famous parachute assault into Objective Rhino in Southern Afghanistan as the first official act of the War on Terror they recited the Creed. 199 combat loaded Rangers bound for a 500 foot parachute drop delivered in unison all six stanzas of the Ranger Creed and upon their return that following morning, as their wounded and dead were lifted from the aircraft and the newly combat qualified Ranger Force assembled battle worn, they did the same.

4

Readying the Battalion: Forts Benning and Stewart

May a dying soldier's last words on the field of battle never be, "If I'd only been better trained.

— Anonymous

It is far more important to be able to hit the target than it is to haggle over who makes a weapon or who pulls a trigger.

— Dwight D. Eisenhower

This is the heart of the book. The main effort. The decisive point in this story. It is where the premise of how the new Battalion was established is told.

Establishing a new unit has happened time and time again in our nation's history. Wars come and go. Leaders peak in heroic fashion and then just often fade away. People forget what was once was and oftentimes history just repeats itself. Salient moments in battle and war are mostly kept for those quiet few genuflecting at cold stone monuments on military posts sprinkled throughout this great land. It's hard pressed to find legacy of any kind being built anywhere, especially in the US Military. However, the establishment of the 1st Ranger Battalion in the post-Vietnam era is not something that anyone in the US Military expected to "grow roots" and not only live on in infamy, but to progress to an enduring capability the country consistently relied upon.

The voice of the founders of the newly formed Rangers takes flight in this chapter as they describe the specificity and details of the newly formed organization. Many understand the basics behind the history of forging of such a force; this chapter unveils some truths yet to be uncovered in the history of the Rangers. Hearing the "real deal" as described by those who were there define this chapter as an exciting focal point of the book.

This portion of the work was initially segmented into three separate chapters. However, the topics of each are so interwoven, so dependent on each other that it made sense to place them all somehow in one chapter. Cadre selection and initial train up of the Battalion at Fort Benning, soldier discipline coupled with fitness and the role of the noncommissioned officers (NCO), combat readiness and performance oriented training (POT) all have dependence on one another in how they affect the establishment of elite force.

Each one of these topics is important. Discipline requires NCO involvement. Fitness requires discipline for the noncommissioned officer must lead teams and squads in this fundamental task. Readiness and performance oriented training are tethered. None of these topics can be achieved without a disciplined force led by quality junior leaders. To achieve success in the initial cadre training required all of these topics executed to the highest standard.

Leuer's priorities in his quest to establish the very first Ranger Battalion since the Darby days of World War II were very deliberate and precise. It was important for Leuer to ensure his Rangers were focused and the priorities he set were decisive. Without focusing on these priorities the elite Ranger Battalion would just become yet another infantry battalion; good at what it does, but not "the best."

Leuer was intent on making his Ranger unit unique and the most capable unit in the Army. To ensure this happened he forged three distinct priorities; 1) Getting good people hired 2) Implementing performance oriented training (POT) and 3) Dealing with "extremes." These extremes are summarized by Leuer as such: "If everyone [in the Army] is running five miles we have to run ten; if they are walking ten miles we have to walk twenty miles. It's got to be done with an attitude that we have to do this to be better and we will be better if we do it with enthusiasm and to standard and hard work."[1]

As an example, when dealing in extremes Leuer wanted to maintain grooming standards that was an order of magnitude more stringent than the rest of the Army. High and tight haircuts or closely shaved heads were the standard for all in his Battalion. To the Rangers this was a badge of honor. Many general officers contacted Leuer about this standard saying it was extreme. However, it was an established standard for all Rangers in his Battalion and Leuer held to it. The high and tight, much like the black beret, became a trademark of a soldier serving in the Rangers. It was something that made them stand above the rest if not just a visible sign of the Ranger discipline. Of note, today's Rangers have loosened their grooming standards, basically because of the requirements to be less noticeable in the special operations environments in which they work in the Middle East conflicts.

CSM (retired) Joe Mattison was one of the very first soldiers in the 1st Ranger Battalion. According to Mattison, many may not know that when they first formed at Fort Benning, the initial cadre consisted of a small handful of Private First Classes (PFC). Mattison recalls that there were about 10-12 privates on the cadre and he feels proud that he was one of them.

When the Battalion first formed, all of the new Rangers were wearing standard, regulation haircuts as outlined in US Army Regulations. No one had any moustaches (they were forbidden in 1/75 from the very beginning) and most if not all, had normal military prescribed hair length.

Mattison was fortunate to be selected to attend the Pathfinder Course as a Private First Class while they were still at Fort Benning. At that time, the Pathfinder Course was five weeks in length, and lower ranking enlisted men and those there on temporary duty (TDY) were required to live in the old Pathfinder Student Barracks on top of Cardiac Hill, just above Lawson Army Airfield (LAAF).

Therefore, Mattison was a 18 year old Private First Class attending the five week Pathfinder Course with a normal, as per prescribed, US Army Regulation Haircut. After a five week absence and graduation he returned back to the 1st Battalion, 75th Ranger area at Harmony Church proudly wearing his brand new, shiny Pathfinder Torch, just knowing that he would get a "heroes" welcome, and hearty congratulations from all who supported sending him to the Pathfinder course.

Mattison walked into the area, and was "attacked" by every noncommissioned officer and officer who recognized him. He ended up doing enough pushups! People were looking at him like they wanted to cut out his heart! At some point in time, during his five week absence in the Pathfinder Course, 1st Battalion adopted the now famous...."Ranger haircut!" – The high and tight.

Mattison was the only one of 1st Battalion, 75th Rangers who still had a regular US Army Haircut, and it was not a good thing, nor a very pleasant experience. He low crawled, high crawled, jumped up and down, screamed and shouted.....and he got his nice and pretty, brand new Pathfinder badge all dusty, dirty and badly scratched. Needless to say, it didn't take him long to make it to the Barber Shop that was just on the other side of the "Ranger Country Store" at Harmony Church, and get a haircut that met the new Ranger standard.

Bill Acebes was his Squad Leader at that time. In his typical fashion, he was the one that "got him out of the line of fire" and out of the "kill zone"!! Not having any money (Private First Class' in those days made $73.00 a month and $55.00 Jump Pay). Acebes threw a $5.00 bill at him and made me get a haircut immediately!!

Leuer also had a distinct philosophy directing his Rangers to adhere to its' principles. Together with the Ranger Creed and Abrams Charter this Battalion would soon establish its foundation as an elite fighting force. Leuer's philosophy demanded that his Rangers first to be gentlemen soldiers, then the toughest, meanest fighters on the battlefield. It was absolutely outlawed to criticize any other unit, branch or service.

As an illustration, the word "leg," a derogatory term often used to describe a soldier who was not airborne qualified, was absolutely forbidden. Where most airborne units freely mocked non-airborne organizations by calling their members, among other things, "legs," the Rangers were not allowed to follow suit. This wasn't just because Leuer wanted to be liked, but rather he knew they would be reliant on others to provide support and absolutely needed their support willingly. With Leuer there was no false bravado with this force, just a "lead by example" mindset.

Leuer was somewhat of a strict disciplinary with the decorum and actions of his officers and senior noncommissioned officers. Unlike in many units, nicknames were limited to the junior enlisted. Names and uniform accessories were his pet peeves. For instance, the C Company Commander, Jerry Barnhill, arrived at a command and staff at Fort Stewart one day wearing a unique C Company scarf. Leuer was clearly not pleased and made a short speech. It never happened again.

Leuer was so meticulous at times that even the wear or practical gear was frowned upon. Chief Warrant Officer Four (retired) Chris Brewer recalls the Battalion rain gear policy. Today soldiers have the most modern Gore-Tex and "poly pro" garments available, and permission to actually wear them. In those days the Rangers had rubber coated parkas and trousers that for some reason always smelled like vomit, no matter how well one thought it was cleaned. Every Ranger was required to carry rain gear but never, under any circumstances, was permitted to actually wear it. They were noisy and had a shine to them when wet, whereas a soaking wet, shivering Ranger doesn't shine and "will let himself choke to death before he makes a noise."

Brewers recalls the first time they ever wore the esteemed rubber suit was much later, when the merciful and benevolent Capt. Jeff Ellis who was the B Company Commander in 1975 unwittingly allowed the donning. It was raining hard and the temperature was as cold as the heart of a Ranger school lane grader. Out of the darkness they saw Ellis stand up in the clandestine patrol base wearing his rubber jacket. They were awe struck, having never before seen anyone dare to actually don the forbidden jacket. A platoon sergeant finally came up to Ellis, and asked, "Sir, can we wear ours?" Ellis looked at him like he was insane, and said, "it's raining, aint it?" Permission was granted down the line to don the treasured vomit smelling rain coats, and jubilation was heard among the ranks. This was followed immediately by squad leaders and platoon sergeants chewing out privates for breaking noise discipline.

Leuer also implemented a zero tolerance policy for disciplinary issues. He demanded that no soldier got into fights at any time or anywhere; unless it was against the enemy in combat. If a public fight did break out and Leuer got wind of it, the Ranger involved in the fight was immediately dismissed from the Battalion. In the end, everywhere the Rangers went they received compliments on how well-mannered and behaved they acted. As the Rangers became more and more combat ready and capable, these respectful Rangers also became universally known as an elite, formidable group of warriors.

The Ranger leaders were both tough and compassionate. CSM (retired) Mike Wagers recalls the first soldier he had to let go. Wagers said that this soldier wasn't all that bad and perhaps he was just not meeting the standards on purpose. On reflection Wagers tends to think that maybe that was his way of leaving. Wagers goes on to explain that it was considered more honorable to get kicked out of the Battalion than it was to just up and quit. In any case, he left and was reassigned to the 24th Division. Two months later he won Division soldier of the month. Even the best of the best were not meant to be Rangers.

By no means did Leuer want to give the impression that his Rangers were soft. Rather he wanted to instill a tough, professional, and no-nonsense leadership character where his leaders would push everyone to question themselves. Leuer was intent to do the simple things perfectly. In doing so, he felt that the complicated things would then become doable. In that respect, he inspired the attitude that even the benign tasks, like drill and ceremony, were as important as live fire exercises. This is why he felt that defining "task" and purpose" for each mission set was vital.

As stated earlier Leuer wanted to teach his leaders to be definitive in their mission sets; not to just "search and destroy" like most of the US Forces were directed to do in Vietnam. Instead, as he puts it, to "go there" to accomplish something like "find that" or destroy a target. By doing this he was talking specifically about defining measurable standards. He felt that measurable feedback was essential in any task. You either hit the target or you didn't; finish the road march on time or not. As a result he implemented performance oriented training (POT). But first Leuer had to assemble the team and organize it so it could train appropriately.

To Leuer there was nothing better than a hard day of work and somebody saying you met the standard and still others saying we did a good job by working hard. To Leuer it was simple – a standard was a standard. Either you said you... hit the target or not... Finish the road march or not... that was critical to a number of things made those selected felt special... And when we get feedback for doing things right it felt more special and we would then step it up to do things right again.

Of note, Leuer talks about this new path that was not yet carved before. However he will be the first to claim he did not do it alone. He had mentors and good ones to boot. Of these mentors Leuer claims that Gen. Bill Rosson – who Leuer served as an aid-de-camp for in Vietnam. Rosson would send him a letter once in a while to see if he would need help and Dave Grange – an icon in the Ranger world, would give him feedback and be open with him. This helped Leuer immensely for the path ahead, while full of successes, would be a bumpy ride.

1 L to r CPT Cubreth Browning Hudson Leuer Gentry Taylor Moses
2 Nelson Birdsall Fowler Nightingale Cobb Skinner Lucus unkn Spain Smith Williams Demoskis
3 Youhouse Bates Fauati Clark Knight Boykins Alderman Aviles Doran Peouarde Dizney
4 McCoY Loley Tetrault Price Rokosz Martin Hendricks Lour Cramer Mattison Champion Brown
Stockwell
5 Conrad Barnhill Schmidt Cobb Reeves Woodfill Kaimi

The original plank holders of the 1st Ranger Battalion. (Source: Nightingale)

The development of the Creed and the constant use of it... these things, if we got them right... you can build a successful unit.

KC Leuer
First Commander of the Modern Day Ranger Battalion

CADRE TRAIN UP – FORT BENNING

The Fort Benning, Georgia phase accomplished several key objectives. It established, on an evolutionary basis, how the Battalion would structure its time both individually and collectively. It created clear appearance and performance standards for the individual Ranger and each subordinate unit in the Battalion at all levels. This critical phase existed to train the cadre on the "what" and "how" of performance oriented training; perhaps the most critical task in all that had to be done. This period was set aside to create and formalize standards as to what the Rangers were and how they would be employed

Leuer wanted to prepare his cadre to conduct this type of standardized training in the intensive training cycle (ITC) he planned once they arrived at their new home in Fort Stewart, Georgia starting in the summer of 1974. A hugely important issue was the necessity for live fire training. It was an area that was moribund in the 1970's Army, but crucial in building the Battalion. Its priority and necessary skill sets had a huge impact on the growth and development of the Rangers from individual level through battalion. It is one of the core legacies of today's Rangers that can be clearly traced to "in the beginning."

The so called Fort Benning Phase, or train up for the cadre for the 1st Battalion, 75th Rangers, ran roughly from March to July 1974. This was an absolutely critical period. In what would become less than a four month span Leuer and his senior enlisted advisor, CSM Gentry, and what they had assembled as a cadre, established the concept for how the command would be established and the schedule of events for the initial training for the cadre. Having personally experienced this type of challenge when the author was chartered to establish the 2nd Battalion, 503rd Infantry (Airborne) in Vicenza, Italy in 2001 it is clear that it is no small feat.

The training plan for the cadre at Fort Benning and then the entire Battalion at Fort Stewart had to be approved by the Infantry Center Command. Once the training plan was approved Leuer was required to brief the Infantry Center's Assistant Commandant, Gen. Richardson, periodically on the progress and issues related to readying the Battalion for its move to Fort Stewart.

Richardson ensured that the Fort Benning staff supported Leuer and the Rangers in attaining personnel, training resources and equipment. The US Army Infantry Center (USAIC) put together a number of classes for the officer cadre. Considering many of the cadre came from the USAIC it was clear that a relationship between the Rangers and Building 4 (the large complex where the Infantry Center was housed that includes classroom and office space) was strong and would last. Today the Ranger Regiment, which is headquartered at Fort Benning, Georgia and the new United States Army Maneuver Center, which is a combination of the Infantry and Armor schools into one center, have a strong bond.

In 1974, Leuer turned to Richardson and his instructors to ensure the Ranger cadre were supported with a variety of requirements to include program of instruction (POI) development, physical training, coordinating live fires, patrolling classes and oversight for training, leadership classes and range coordination for marksmanship and field firing. In that it was summer at Fort Benning the Rangers appreciated the air conditioning in Building 4; a creature comfort that was not commonplace at the time throughout the US Army facilities. What Richardson provided mostly was guidance and advice to Leuer as the commander's advocacy for the Rangers. The effort to synchronize the effort to resource the Battalion became more and more organized as Leuer's staff came on board and were able to work staff actions via the various staff elements on Fort Benning and elsewhere. The integration of a viable staff into the process relieved Leuer from the need to get involved in all staff actions allowing him to focus on command related issues.

This period, as expected, was extremely busy. The small Ranger command and cadre had to mesh training, team building, recruiting trips, and advance work with Fort Stewart so when the Rangers conducted a parachute jump there on July 1, 1974 they would be ready to start their intensive training period. The goal was to assemble the staff, a cadre of the subordinate units and gather the necessary equipment so that the 1st Battalion would look more and more like an organized unit prior to their move to Stewart.

Leuer personally selected the initial cadre at Fort Benning which consisted of a total of 51 men. Most of the members of the original Fort Benning cadre were Sergeants First Class, or E7, and above in rank with about 90 junior soldiers. Training this cadre became the task at hand. Being located at Fort Benning during the initial standup gave the new Rangers great opportunity. Leuer and his Rangers could take advantage of the facilities available to attend specialized schools such as Airborne, Jumpmaster School, Pathfinder School, and Sniper School.

The Four first Ranger Company Commanders, Rokosz, Barnhill, Clark and Nightingale.
(Source: Currie)

Jerry Barnhill, the original C Company Commander recalled that all officers and line Sergeant's First Class (E7) and above were Ranger qualified. One of the requirements for selection of officer and noncommissioned officers to the Battalion was that they had to have the Ranger Tab – or be Ranger qualified. This meant that they attended the nine week school held at the Harmony Church area at Fort Benning and if they graduated were awarded the black and gold Ranger Tab. This was no small feat for the attrition rate in Ranger School hovered around 65% at any given time.[2]

When selecting lieutenants from across the Army, Leuer shared the respective lieutenant candidate folders with each of his company commanders (Clark, Rokosz, Barnhill, Nightingale) in a meeting in the unit's mess hall. There each commander reviewed the files and then made recommendations to their battalion commander as to what lieutenant would go where in the Battalion based on personalities and experiences.

But manning the Battalion with the right people was just one task of the many placed on Leuer in developing this new unit. Certainly there was past precedence since Ranger units would come and go for centuries. The goal now for Leuer was to define this unit in such a way that the effort put into making it an elite force, not only capable and needed in our military arsenal, but to make it so the first time in our nation's military history that the Ranger unit would persevere in the ranks.

The Fort Benning Infantry Center's force developers designed the first Ranger battalion to be a very light organization. Leuer recalls that it was approximately 588 men with a very small headquarters that included of about 50 men. There were three rifle companies each with a weapons platoon that had 60 mm mortars organic to the company. 60 mm mortars are hand carried (three to four man crews) and shot in the conventional mode or in the handheld mode.[3] The heaviest weapons originally in the Battalion were the 60mm mortars and the 90mm recoilless rifle (RCLR); a bazooka type weapon system first seen in some form during World War I.[4]

The Ranger equipment and gear initially came from a combination of Benning and Forces Command stocks. Nightingale recalls that the 90 mm recoilless rifles came from depot stocks and storage and the M16 rifles were new. While some of this equipment came to the Battalion either new or newly refurbished some of the mission essential equipment was quite old. The M60 machineguns were more recent models most of which were used in Vietnam. The mortars were of World War II vintage. The uniforms issued to the Rangers came from the Republic of Vietnam stocks. So like any new unit or organization it was piecemealed together from various locations and through various efforts.

Even though they were a light unit, the ability to organically move Rangers with a wheeled capacity was almost nonexistent. The only two vehicles in the Battalion were two M151 Utility truck vehicles in the Battalion Headquarters. No other units in the Battalion had vehicles. The M151 was commonly referred to as the "Jeep" which was a four man wheeled vehicle.

Other than these two vehicles to help move people and equipment around, the Battalion had absolutely no organic sustainment units. Therefore any supplies the Rangers needed in the fight would have to be carried on their backs. This logistics shortfall proved challenging for the early Ranger units. The Ranger logistics capability would evolve over the next few decades with the introduction of the Ranger Support Elements (RSE) that supports home station training and the Joint Special Operations Command (JSOC) which produced integrated sustainment support for operations for the Rangers and other special mission units (SMU).[5] But that would be in the future and for now the Rangers were left to fend for themselves.

Keith Nightingale sums up the challenges the new Ranger cadre experienced. Besides the ongoing challenges to recruit, Nightingale recalls that the command had to literally define what the Ranger Battalion stood for, how it should be constructed, how it should be trained and by what principles should the training methodologies follow. Once these questions were analyzed and answered then the command could decide how it could and should be employed.

Defining the Ranger Battalion's mission essential tasks and combat roles were dependent upon not only the capabilities of the Rangers, but also upon the wartime contingency plans and requirements that existed globally. The process of integrating the use of this expeditionary force into the multiple combatant commands war plans was no small task.

Nightingale describes the process of developing the Ranger's wartime requirements as iterative. Working with the 18th Airborne Corps and the Joint Exercises Branch of the Joint Chiefs of Staff (JCS) office, as well as Forces Command (FORSCOM), the Readiness Command (REDCOM) and the US Army Training and Doctrine Command (TRADOC) was critical. One of the most critical tasks and also the most daunting was getting these different commands educated on what the Rangers did and how they should be employed in as part of the various contingency plans (CONPLAN) and operations plans (OPLAN). Getting an entire military in synch with Ranger capabilities and how they were employed was a great challenge.

Not only was it difficult getting these organizations to understand the purpose of the Ranger Battalion, it was a shock to all as to the amount of logistics that was required to sustain them. The Ranger planners would go to the various headquarters and show them their table of organization and equipment (TOE). Those that knew little about what the Rangers did would look at the TOE with surprise. There is no doubt that defining how the Battalion would be supported logistically was a monumental effort. The Joint Staff just couldn't get past the idea that the Rangers were not just a light infantry unit that should be used conventionally instead of as a special operations unit in a quick strike role.

The Rangers wanted to ensure they could convey their cause by speaking articulately about what it is they were capable of performing in battle. Relating to their heritage, they did a lot of reading regarding earlier Rangers in the likes of Mosby, Rogers, and Darby in an effort to select not only key, but highly successful points from history they could and would integrate into their effort. There was a lot of collaboration occurring at this time to ensure the Rangers could market their capabilities, to the various combatant commands.

Initially, commands simply just did not appreciate having to pay the logistics bill for Ranger participation in their exercises. And if they agreed to use Rangers in their exercises, the secondary issue was inappropriate employment. There was a tendency to suggest using Rangers as a super battalion or mini-brigade in a conventional infantry setting vice strike operations.

There was one incident at Forces Command when a colonel in the plans section said "If the Rangers won't do what we want them to do, we just won't use them." Nightingale recalls thinking that, "I think Lt. Col. Leuer had to 'adjust' the mindset. " These big picture challenges were going on at the highest levels in the Army and Joint community. At the same time, Leuer was honing in on soldiers specific standards for he knew that there was no doubt it would matter not what a combatant commander wanted out of his Rangers unless they were highly trained and highly disciplined to get the job done right.

UNWAVERING STANDARDS AND BASICS

CSM (retired) Mike Wagers, a C Company platoon sergeant in 1974, said that Leuer explained to each Ranger in his command that they would be training "to standard." He explained that they would conduct physical training to standard, field strip a weapon to standard, and cross a linear danger area to standard. Everything they would do would be conducted to standard.

He goes on to state that he is not sure what the Rangers thought, but he for one did not expect what they got. You name the task and they had a standard for it. You name the event and they had a standard. Those standards had been analyzed, scrutinized, debated, researched, tested, then tested again and finally written down to ensure they were correct. How long did it take to field strip a weapon and reassemble it? How long should it take in the dark and in the rain? How long did it take to wake up, roll up your sleeping bag, get your rucksack packed and put on your back and be kneeling on one knee in movement formation ready to move out at three o clock in the morning? He summarizes by saying, "I am not talking about making swag or estimated about how long it takes; I am talking about KNOWING EXACTLY."

What impressed CSM (retired) Mike Martin the most, when he served as the original First Sergeant (1SG) of Company B during the period of its activation, was the individual discipline. He applauds both the personal discipline of those soldiers who lived the Creed and respect those leaders that demanded such a high level of discipline and unwavering standards.

Martin, an inductee to the Ranger Hall of Fame, went on to serve at the highest levels in the Army culminating a career that he began as a smoke jumper and eventually retiring as the most senior enlisted man in the Republic of Korea. To Martin, there was a sense of integrity in demanding compliance with standards. Most importantly Martin was consistently impressed by the positive attitude of the junior enlisted men who were capable of taking initiative on their own.

Col. Dave LeMauk was a young private first class radio telephone operator (RTO) at the time the 1st Ranger Battalion was formed. To him serving in the Battalion was the most positive and enduring influence in his life, although he didn't realize it at the time. It was his first assignment, and he grew up in the Battalion assuming the rest of the Army was just like that (or at least relatively close). Only later did he learn that the Battalion's standards, whether physical, mental, or intellectual, were far above what other units in the Army could demand.

LeMauk did what his noncommissioned officer and officers asked, and never thought twice about it. After a spending a year in the Battalion, he went to Ranger School with some of the other soldiers in the platoon where they joked about taking a break from the physical grind of the Battalion. Ranger School was tough, but in many ways the Battalion was tougher.

He recalls they often went on long runs, and usually at a pretty good pace. Most new folks had trouble keeping up, so they instituted a special physical training program for the first 30 days. After that, everyone was expected to keep up with their unit.

There was a great deal of encouragement given to keep Rangers in the formation, and they expected everyone to give 100%. But once you gave up and fell out, that was it – no one wanted quitters in the Battalion. Soldiers who fell out were greeted with jeers and physical abuse as following units passed them.

While physical fitness training was excellent, it challenged the older noncommissioned officers. For starters, one of the goals of the Rangers was to be the fittest unit in the Army. To accomplish a task like mastering physical fitness meant tough but appropriate physical training with variations that built espirit-de-corps, while building strength and endurance. But to do this you had to first be disciplined.

Ranger haircuts and uniforms would stand out. Road marches were essential building blocks. Rangers had to move fast and far on the ground and that took skill, conditioning and repetition. Training the mind was equal or greater a task than training the body. Drill and Ceremonies (D&C) would be an integral part of looking and acting elite. The Rangers would learn how to march in formation and how to give and take commands. Every physical training event finished with D&C as the cool down technique. Everyone, regardless of rank, learned to "give" D&C to the group. It was events like this; Leuer realized early on, that was building his cadre so that they would accept high standards.

Leuer said that attrition rates were heavy in the first few months because it was difficult for many to maintain the high physical standards. For the Rangers, physical training wasn't just running and pushups. It was combat related physical challenges with the likes of road marching becoming a trademark for the US Army Ranger.

Martin remembered the Battalion deployment to Dahlonega, Georgia at the end of the Benning phase just prior to the parachute jump into Fort Stewart. There the Rangers conducted raid and ambush operations, culminating in a 25 mile road march. First Sergeant Martin knew the area and helped Major Philip Browning, the original Battalion S3 and today a retired Brigadier General, plan the route.

There was a stringent time standard for finishing the road march. Leuer was unforgiving in failure and recalls stating that if we missed it by a minute; we would repeat the march the next night. He understood intent of focusing on the basics. Staying in step in physical training formation run or doing close order drill was just as important to Leuer as setting up an ambush that has a 100% kill. Everything his Rangers did he wanted it to be a part of perfection and to standard. He constantly questions his subordinate leaders on how hard do they have to push to convince everyone "you gotta be to be 100% to win... can't be 90% and thinking you can be better if you work at it you have to believe that we are as good as we can be for we gave it everything we possibly could give it. And that is the right ammo and the right feedback."

Road marches and physical training runs were conducted to standard with five to ten mile runs and five to thirty mile road marches sprinkled throughout the training schedule. Specialist Fourth Class (SP4) Frank Magaña of B Company, who eventually would rise to Ranger fame as the CSM for 2nd Ranger Battalion, said the road marching was incomparable to anything else he had ever experienced. The most difficult standard being the 5 mile in one hour standard, with all table of organization and equipment (TOE).

Junior leaders like Martin were impressed with Leuer's focus. Martin was also impressed with the level of emphasis on marksmanship. There is no doubt that the Rangers had to master their shooting skills. He and other Rangers agreed with Leuer that "shooting" was important and, to a man, they were determined to get his Rangers to support the commander's intent in that area.

In field exercises, the standard of having to sterilize the area was critical indicating the discipline of the Ranger even in rough field conditions was unwavering. Todd Currie commented that it was essential for junior enlisted to learn basic skills to perfection. This type of training and mindset would form the foundation for the unit for the future.

Martin believed that in a way to instill discipline would be to have a greater emphasis on martial arts and stealth skills. That desire finally took hold for today's Rangers all have hand to hand jujitsu training making young Rangers confident and tough. Coupling the demand for high standards was a zero tolerance policy for behavior in accordance with the Uniform Code of Military Justice (UCMJ) and policy set forth by the various commands. This was a time when all knew that the Army was recovering from an environment where discipline was all but missing from the formation.

Rokosz tells a story relating to this type of unwavering discipline as follows: The battalion gathered in formation at attention in front of the barracks. There were Marines vacating the barracks at Stewart when the Rangers arrived. The Battalion stood at attention with not a muscle moving an inch as the Marines, in an ill disciplined way, poured beer out of the upper floor windows on the Rangers heads. Leuer arrived at that moment and saw what was happening. Beaming with pride at the disciplined displayed by his Rangers and fuming with anger for the Marines who defaced his command, Leuer hunted the Marine Commander down as the Rangers remained at attention, soaked to the bone in beer but unwavering in their self-control.

Standards, discipline and realistic training became the three anchors in the Battalion as its existence as a fighting force became a reality.

Standards and discipline facilitated an ability to train realistically. Realistic training became the hallmark of what it meant to be a Ranger. Extensive use of live fires and sand tables in mission planning as well as rehearsals became standard to all training events. But first Leuer and his leaders knew they had to rely on quality noncommissioned officers to get the job done.

THE RANGER NONCOMMISSIONED OFFICER

Ranger, you are assigned to Battle Company, 3rd Platoon. Your unit is three thousand five hundred meters from your present location on a magnetic azimuth of 310 degrees. Take charge of this patrol and move out!!
Instructions barked at Private First Class Chris Brewer when asked who the senior man in the back of a truck full of privates was

The role of the noncommissioned officer was perhaps the most crucial component in making the battalion. With the high expectations placed on the 1st Battalion, the leadership was under great pressure to produce results. History shows that the noncommissioned officers can make or break an organization.

According to Nightingale, training in the Army at that time was in abysmal shape. Probably more of a reflection of the Army at the time; an officer was afraid to assert themselves, there was little distinction between officer and NCO roles, and the Army lacked a coherent philosophy regarding either training or leadership and the combination of the two.

This necessitated a major change in the training methodology at the small unit level. The Vietnam experience gutted the Army of its experienced leaders. Those that enlisted in WW II and Korea, paid the price and learned the hard lessons of combat, but stuck it out in the early Cold War. Over the course of the thousand day war, Vietnam lost many of the experienced leaders as casualties, retirees or cop-outs. Many of these warriors simply grew tired of trying to do their job with minimal support and eventually quit trying.

The officer corps grew risk adverse, personally self-centered and lost its professional way. The core competency that was so apparent on the beaches at Normandy did not exist in the Mekong Valley or the many base camps throughout Vietnam. The difference between a company commander, a private and a squad leader or platoon sergeant became very muddy and ambiguous. The definition of roles was lost in the process. This is the Army Abrams sought to fix with the inclusion and infusion of the first Ranger Battalion. It was time for change and leadership responsibilities were forefront in the need for change. Everyone had a job and everyone had to know their job.

The selection of noncommissioned officers in the Battalion had to pass the scrutiny of the cadre of first sergeants and platoon sergeants assembled in the early stages of the organization. These select noncommissioned officers had been through everything from day one and were extremely judgmental and critical of other noncommissioned officers introduced to the Battalion especially noncommissioned officers that "couldn't hack it."

CSM Wagers recalls that it was a tremendous sense of pride being one of the original platoon sergeants in the 1st Ranger Battalion. There were nine Ranger Rifle Platoons in the whole Army and you were a platoon sergeant of one of them. Nine! This kind of excitement and attitude wove itself into the core of the original plankholders.

Being a noncommissioned officer in a Ranger Battalion was not an easy living in 1974 and isn't today. But it's the pride and adherence to standards that makes it special. 1SG (retired) Steve Rondeau recalls that he had been in the unit for a year and was selected to appear before the E-5/E-6 Ranger promotion board. The promotion board was a two-day process consisting of one day of "hands-on" evaluation beginning with physical training. This was followed by additional physical demonstration of noncommissioned officer job skills. The second day was a board appearance to assess the candidate's leadership knowledge base.

Maj. Downing was the board president for Rondeau's board. CSM Henry Cairo and the unit first sergeants conducted the actual board and scoring. Most officers treated promotion boards as noncommissioned officer business and didn't participate except to provide a signature after the board was completed. Maj. Downing did not hold with this philosophy.

The large number of candidates during this particular board caused the PT portion to run past its planned duration. After the last man had finished leading his designated exercise, straight from the Field Manual, Cairo looked at his watch and announced that there was no time for a run. He was immediately interrupted by Downing who said in a loud and thunderous voice, "Bullshit Sergeant Major, There is always time for a run." Rondeau looked at the first sergeants and could see a look of consternation on their faces. Maj. Downing then took charge of the formation, moved them to the cross street and took off.

As Rondeau puts it, there was no stretching or touchy-feely Master Fitness Trainer BS in those days. The Rangers wore jungle boots and were desperately trying to run in an orderly formation behind a man that could run a sub six-minute mile. By the pace Downing was setting, which was quick for a long run, Rondeau estimated it would just be a two mile run.

To his chagrin, they hit the wood line and started down the sand trail. By now the first of the noncommissioned officer candidates began to fall out. The pace had not slowed an iota. Rondeau thought to himself "surely he'll turn at the dump and make it a 3 mile run."

The candidates continued by the dump and more Rangers fell out. A few more were starting to vomit. They then turned down the sand trail that led to the ammunition area dump, meaning this was going to be a long run. Rondeau recalls experiencing a moment of panic. This was closely followed by the words PSG Joe Alderman used to tell them in times of trial. "The mind quits before the body, Ranger." Remembering what Alderman said motivated Rondeau and he continued to run.

When the candidates finally returned to the Battalion area the formation was much smaller than when it started. They had just ran five miles in thirty-two minutes. Rondeau thinks he made it because of Capt. Don Clark's rigorous PT regime in A Company, SGT Bill Wallace running beside him and motivating him the whole way, and Joe Alderman's sage words repeating themselves in his head.

Downing, in his signature command voice, told Cairo to collect the fall-outs. When everybody had been gathered up Downing told the stragglers they were not in good enough physical condition to be combat leaders and would not be appearing before that promotion board. As a result of that dose of reality, several men terminated Ranger status that day and were immediately re-assigned to the 24th Infantry Division.

There was little slack when teaching and coaching. The new noncommissioned officers either got it right or quickly moved on. Many were simply defeated after a short time in this process. Ultimately what happened was that the junior soldiers took over through attrition. The key to making the modern day Ranger Battalion work was making the noncommissioned officer structure work. The respect once deeply imbedded in the noncommissioned officer corps was all but demolished in Vietnam and not resurrected in the post-war Army. Leuer knew this had to be fixed or the concept of a highly disciplined and highly special operations unit and masterful light infantry unit just wouldn't work.

The noncommissioned officers' byproducts of the Vietnam War and the attitudes, lack of discipline and years of physical neglect was a hard thing to overcome. Many noncommissioned officers, no matter how good they were perceived, just couldn't overcome the legacy that years of experience in the Vietnam era saddled them with. Therefore a young specialist would become the acting squad leader. As time went on, these young soldiers would become seasoned leaders with the instilled leadership qualities and physical capacities demanded of Leuer and his Ranger leaders. What became "growing their own," or grooming young soldiers to become noncommissioned officers in the Ranger units bore its roots during the initial stand up of the 1st Ranger Battalion.

Despite intensive recruiting, many noncommissioned officers arrived incapable of assuming responsibility or were too selfish to make the necessary sacrifices for the organization. Most of the noncommissioned officers recruited, according to Nightingale, were simply not prepared to pay the physical price or bear the leadership burden. They really just couldn't cut the mustard but were so mesmerized by the "glory" and the opportunity for a black beret so they jumped at the chance to join.

In the end, they either quit or were compelled to leave by peer and subordinate pressure. Growing their own was much a reaction to necessity as to acknowledgement that only exposure to and success in the emerging battalion could prepare a soldier for increased responsibility within the unit. The dynamic, intelligent Private First Class (PFC) or Specialist (SP4) became the squad leader and the high performing squad leader became the platoon sergeant. They did the work and they worked to a standard where the older, more experienced noncommissioned officer was incapable of inculcating success. Because many of the mid-level noncommissioned officers could not maintain the physical standards, the junior enlisted in many cases ran the noncommissioned officers into the ground causing them to quit out of embarrassment. Very few recruited noncommissioned officers had to be asked to leave for non-performance. Their subordinates made them leave.

It was no doubt that units at each level succeeded or failed largely due to noncommissioned officer competency. Making the noncommissioned officer responsible and accountable was a major, if not the major, aspect of Leuer's tenure. The attrition rate for platoon sergeants and squad leaders was extremely high-probably in excess of 200% of assigned. Many simply did not want or were unable to carry the burden of direct responsibility for subordinate performance. CSM Don Bruce, a platoon sergeant in B Company in 1974, said that the squad leaders formed the basis of the Battalion for setting and enforcing standards, discipline and unit cohesion. It was critical to have the best squad leaders possible. Growing their own was the way to go and Leuer knew it.

Col. (retired) Jim Moeller was a young private in the 1st Ranger Battalion right after it was formed. His first Ranger assignment as a private in B Company's 3rd Platoon as a member of 3rd Squad. Two years later he was the squad leader. That is how the Rangers "grew their own." In a lot of ways it was *safer* for the Rangers to groom their young, instilling the standards and then making them leaders in an organization they understood.

Leuer had the wherewithal to see that while implementing tough realistic training was an important element to proliferate throughout the entire Army via the Abrams Charter, implementing discipline in the entire force was as equally paramount and no training of any sort could realistically take place unless the unit was disciplined.

Quality among the platoon sergeants and squad leaders was problematic. Even more problematic was the aspect of heart. Being a Ranger in the emerging performance standards required a great deal of self-discipline, goal-orientation and sense of responsibility. This was particularly true of the platoon sergeants and squad leaders as their successes or failures were clearly public in the form of the troops in front of them. There was no place to hide.

The eventual integration of the performance oriented training methodology put the direct burden on the noncommissioned officers to maintain but they were fully unprepared to accept responsibility. A number of noncommissioned officers chose not to volunteer, as it was clear to them that the demands on their time were more than they cared to dedicate. Staff Sergeant Steve Hawk, a Private First Class at the time, recalls that a lot of these noncommissioned officers just came back from Vietnam and they weren't going to spit-shine a pair of jungle books. It was a culture shock to many of these "old timers."

According to Nightingale, this was especially true with Special Forces noncommissioned officers. Too many in the Special Forces community at this time made it clear that their family situation would not support a Ranger-type environment or, more realistically, they had a pretty good situation at present and didn't want to jeopardize their career paths risking a transfer to the Rangers. Regardless of the shortfalls, the Fort Benning cadre were first class soldiers, in fact, the best the Army had as judged by Leuer and Gentry.

Regardless of the challenge, there were clear examples of those noncommissioned officers who rose to the challenges by embracing the standards and performing brilliantly. SP4 Frank Magaña of B Company, a future Command Sergeant Major for 2nd Ranger Battalion, went to airborne school enroute to Ft Stewart, arriving in early July, 1974. He stayed in the battalion until Feb 1977, leaving as a sergeant-promotable (SGT (P)). He eventually served as a platoon sergeant with the 2nd Ranger battalion at Fort Lewis, Washington from 1982-1985 and in 1983 participated in Operation Urgent Fury in Grenada. He then served as the Command Sergeant Major for the 7th Ranger Training Battalion which ran the desert phase of Ranger school at Fort Bliss, Texas.

In his first tour with the Rangers, Magaña was assigned to B Company. He said the biggest influence on him was his squad leader, SSG William Acebes who would eventually retire as a Command Sergeant Major. Magaña felt that learning from Acebes had a major influence on his entire career. Acebes was calm, yet firm, extremely competent, and treated all squad members with respect. He personally helped Magaña prepare for Ranger School which he attended some six months after arriving at Stewart. Acebes prepared him so well that Magaña was the Distinguished Honor Graduate of his course and was the winner of the Merrill's Marauders Award. Given the strenuous training regimen Magaña experienced at Stewart during his train up he felt he was more than prepared for Ranger School and it showed.

Magaña felt that Ranger noncommissioned officers really began to earn their money after leaving the battalion and assigned to units throughout the Army where people expected them to know everything. He emulated Acebes style of leadership and understood and embraced the battalion training process and applied these throughout his career. Magaña feels that the pride and esprit he experienced in the 1st Rangers could not be matched anywhere. It inspired and motivated him to serve in the Army for 23 years. He was selected and inducted into the Ranger Hall of Fame in 2004. This is one example of how a great Ranger noncommissioned officer could have a wide ranging impact on, not only an extremely successful noncommissioned officer, but a great part of the US Army; it alone was one legacy the 1st Ranger Battalion provided.

Another example of a noncommissioned officer who rose the occasion in the early days of the Battalion was the personnel noncommissioned officer in charge (NCOIC), SFC Joe Spain, a tough minded senior noncommissioned officer from Tennessee who, according to Leuer and Nightingale, was a "real miracle worker" working long hours to support the command and accomplish what others struggled to get done. At a time when it was tough not only to recruit but also to overcome the many levels of bureaucracy, Spain was the "go to" guy to figure it out. According to many of the leaders at the time, he was, outside of Leuer, the most significant member of the cadre team.

Rangers rising to the occasion and excelling became a staple in the Regiment for years to come. What impressed Jerry Barnhill greatly when he commanded C Company was that the officers and noncommissioned officer worked hard together, as a team capitalizing as a whole with Rangers individually excelling, to figure out best ways to make training performance oriented and led by great NCOs. But let there be no mistake in understanding, as Mike Wagers stated when speaking of the discipline of this new unit, "If there was a holy grail in the 1st Ranger Battalion at that time it was the standards" This discipline led to a methodical training regimen. This new training methodology – a system that measures proficiency – which still exists today even in a more expanded way, had a significant impact on every Ranger entering the Battalion at that time and an even larger impact on the rest of the Army in the years to come. Measuring success for these new Rangers began in the most rudimentary form, on the physical fitness field every morning.

PHYSICAL TRAINING

Another part of training was physical training and athletics. Leuer used morning physical training (PT) starting around 0430 as a teaching and training vehicle for everyone from lieutenant colonel to private. In the beginning he personally administered PT to his cadre, establishing the standards for its conduct and then ran as a unit with him in the lead. It was difficult to sustain but not impossible and drew everyone together as a shared experience. He makes the comparison of the now famous Band of Brothers episode running Currahee Mountain in which the leader led a rigorous session and the soldiers motivationally followed. As a side note, Leuer ran PT runs with his dog Daisy. Soon her attendance became commonplace. The mantra became if the "old man's" middle aged sheepdog/retriever breed four legged friend can make the runs than every Ranger should be able to too!

Physical training in the 1st Ranger Battalion didn't only consist of doing calisthenics and running. Leuer ensured that his Rangers had, as he called it, varied PT. Besides running a laboring distance he implemented station training and organized athletics. Not only did this intense PT program make his Rangers fit, it drew them together, exposed their weaknesses and demonstrated the variety of options. One of their final acts at Fort Benning, Georgia and as a means to validate the tough, rigorous PT regimen implemented by Leuer was an extended station run where it was performed as pure competition. Graders were at each station to insure compliance and his Rangers embraced the competition and performed as best they could regardless of rank or position.

With a challenging PT program comes a personal involvement by its leaders at all levels below the battalion command. CSM (retired) George Horvath who was the first platoon sergeant in B Company had been recently selected to Master Sergeant (MSG) and was made the First Sergeant of Headquarters and Headquarters Company. He was given the task of selecting the route for the Headquarters and Headquarters Company annual five mile certification road march. Horvath felt confident he chose a good, solid route.

Horvath took off that fateful evening with the Battalion Executive Officer, Maj. Rock Hudson in the formation and Horvath set the pace. The group was having a difficult time making time hacks at each mile marker and so they kept trying to pick up the pace; hollering, yelling, shoving and screaming at each other like Rangers do! But in the end the group failed to meet the goal of five miles in one hour. After they crossed the finish line, Headquarters and Headquarters Company was spread out along the road, panting, moaning, groaning and looking like they had just fought our way out of a Viet Cong ambush. It wasn't a pretty sight.

Hudson was livid, he went out and re-measured the distance and found it to be nearly six miles long, not five! So he comes stomping into Horvath's office "pissed to the max" and demanded to know how he could have messed the distance up so badly. Then Horvath remembered he used his pickup truck that had recently been fitted with oversized mud tires and failed to get the odometer re-calibrated. When they did the math, they had in reality set a Headquarters and Headquarters Company record for five miles, but "Rock" would not let that stand and he insisted that they do it over again with the route he had personally measured.

Don Bruce recalls a specialist fourth class (SP4) named Elesa who was well liked, but was struggling to meet the physical standards. As his noncommissioned officer, Bruce decided to work with Elesa nightly, taking him through extra physical training. To his surprise, the entire squad showed up and took it on as a squad mission to help Elesa make the standard. This effort took no prodding from anyone for the squad felt compelled to take ownership as a team.

Elesa eventually met the standards. Later Elesa asked Bruce for help on a 20 mile charity run in Savannah. Elesa had a group of 20 enlisted men who wanted to run the event, but Leuer wouldn't approve it unless a noncommissioned officer was with the Ranger enlisted men. Bruce volunteered to accompany them and the group ran the 20 miles in formation and finished 9th overall. Leuer was so impressed; he asked all officers donate to the charity.

It is obvious that teamwork meant something to the Rangers and teamwork was critical to success in combat. Surviving in combat was the goal. Achieving a high level of physical ability was the vehicle to achieve that goal. CPL. Jeff Everett was an assistant M60 Machinegun Gunner when he was training with his gunner, SPC Watkins one hot summer day in 1974 at Fort Stewart, Georgia. The mission that day was to practice live fire attacks where they would move through a knee high marsh as a team. It was an uneven run and it had to done quickly. Running in a swamp in combat gear is not an easy task. Suddenly, Watkins went face down into the water weapon in hand. Everett pulled him up and quickly realized he was out cold. They halted the attack and then pulled back to starting line. Everett's team carried Watkins out of the marsh, and propped him up against a tree in the shade, pulled his steel pot off and tried to bring him around.

Watkins was a fellow Ranger, and a damn good one. He was awake, but unable to speak, he could mumble, but not pronounce a clear word; his brain had been fried from the heat and amplified by the difficult move. The team stayed with him cooling him down, but Watkins couldn't actually pronounce a word clearly for many hours. This event validated the Rangers hard work realizing that was what their training was really all about, taking one's self to the limit, becoming a Ranger, a fighting force on the field of battle that would be undefeatable. It was learning just how much the body and mind could be subjected to and still complete the mission even if you were the lone survivor. It was not memorizing the Ranger Creed; it was the ability to live it.

Another Ranger that embodied the physical aspects of the Ranger Creed was Todd Currie. Currie arrived at Fort Stewart in August 1974 fresh out of basic training at Fort Dix, New

Jersey. He attended Infantry Basic Training at Fort Polk, Louisiana, and the Airborne School at Fort Benning where he was promised a Ranger slot and then go to Fort Bragg, North Carolina to serve, as he puts it, with the "82nd Airplane Gang." A Ranger recruiter has such an impression on Currie and his buddy, Melvin Irving, while they were in jump school that both men found themselves with the Rangers shortly after graduation.

To Currie, physical training was the hallmark for Ranger success. He recalls a twenty mile road march where he was assigned as the ammunition bearer for the 90 mm recoilless rifle. During the march the team members would rotate the 90mm between each person in the section. Currie remembers the others falling out and he was left stuck with carrying the nearly 40 pound rifle the last ten miles.

At the breaks, his company commander Ron Rokosz asked Currie's platoon leader if he was going to make it. Bill Caldwell, a retired three start general, told him he would because Currie wanted to be promoted to private first class (PFC) before Christmas. Currie finished the march and then passed out. Revived at some point later, Rokosz shook his hand for making the tough standard; but also told him he was not getting promoted that day. This just made Currie work harder and he eventually made the mark.

Currie said that the physical training was very demanding and there was incredible peer pressure, much like there was in reciting the Ranger Creed where if one made an error it was a mark of failure. He recalls that the physical training regimen was tailored to fit the upcoming missions by working the muscles needing the attention of the mission set. Today's Army's combat focused training regimen foundation is a product of this early era Ranger PT program.

COMBAT READINESS AND IMPLEMENTING PERFORMANCE ORIENTED TRAINING

This portion of the chapter focuses on the events that took place immediately after the formation of the force. Early training deployments; most notably to Fort Stewart, Georgia and other training conducted in the early years that set the path for success. The establishment of performance oriented training (POT) methodologies introduced the high standards for realistic, tough training and the combat focused realistic approach to the Army as a whole. This is important for the execution of the Charter was critical in the acceptance of the force as an elite trained and ready arm in the Army.

While still at Fort Benning one of the very first things Leuer did was introduce the cadre to how the Battalion would implement the performance oriented training he learned at Florida State in the previous year and a construct the rest of the Army was slowly implementing. The course he attended at Florida State proved fruitful for he also learned how to teach this training regimen. Leuer saw it as the primary teaching and training methodology for the Rangers. The Ranger's would be the first to introduce this to the Army and to take it on as the base methodology for all Army training management.

The point of all this was to make training become "training to a standard." A standard was a clearly measurable, quantifiable demonstration of performance in a specific task or subset. There were several reasons why this was important.

Historical Army standards, or more accurately, goals, were very subjective in nature and historically dominated by the decision of the senior man present. Querying if a unit perform or not perform "satisfactorily" was always the question commanders wanted answered in training and in combat.[6] It is hard to teach and train to that question as everyone had to know what "satisfactory" meant to the grader. The Army Training Tests (ATT) and Operational Readiness Training Tests (ORTT) were never able to successfully define what "success" meant for they were used subjectively against a so-called ladder of achievement for a unit of a specific Table of Organization and Equipment (TOE) should be able to perform.[7] For both of these tests the unit being tested really didn't know the benchmarks for success. Leuer was given the task to devise a more measured way of doing business.

Part of the cadre training was to dissect the existing Army doctrine and publications regarding individual and unit readiness training. Trying to turn those documents into a specific usable, applicable and meaningful training programs that delineated responsibilities and achievements was at the best difficult.

To add to this, Leuer volunteered for and was awarded the task of writing the first Army Training and Evaluation Program (ARTEP). It was a great opportunity for he could develop the newly forming Battalion at the same time infusing this performance oriented training and evaluation program as a test case. POT would be used as the base and outline how an infantry organization would go from individual to battalion level skills and be graded on the same.

Jim Montano, the S5 or plans officer was tasked with the heavy lifting in this task. Each unit to include Headquarters and Headquarters Company (HHC) and the line companies (A, B and C) would be responsible for developing the Task, Condition and Standard for each appropriate component.

Over the course of his tenure with the Rangers, Montano developed two distinct training related items that proved to be an integral part to the performance oriented training concept that Leuer had championed. Montano was responsible for the development of the Ranger Indoctrination Program (RIP) that ensures all new Rangers can meet a concise list of performance measured tasks and the development of relationship POT and ARTEP at squad, platoon and company. As Montano developed lesson plans and training scenarios there was a big emphasis on night operations stressing mental and physical fitness. Things like how to employ a machine gun, treat wounded, call for fire, maneuver a platoon, etc. were all task requiring a focused training plan. Developing measurable standards for topics like these would be the foundation of everything else.

Mission planning, rehearsals, parachute insertion, assembly procedures, attack, defense, ambush, raid, river crossing, patrolling, movement to contact, movement techniques, contact drills, road marches and live fire tasks were all on the table for formulating into measurable lesson and training plans. According to Montano, once these training plans were built, the Leuer period focused on basics at the squad and platoon level as building blocks. The years after Leuer and with the inclusion of the 2nd Ranger Battalion into the force, allowed for progression in the training management process. As time went by there was a natural shift to company and battalion synchronized and combined operations as well as airfield takedowns and no-notice exercises. All this eventually led to the beginning and refinement of special and counter-terrorist operations. However in 1974 Leuer and his Rangers stuck to the basics.

The Army Test and Evaluation Program and Performance Oriented Standards were interrelated and key to measuring the success of all the Rangers during this period. Leuer brought in the cadre to write the basics of a light infantry battalion from individual to battalion level. Todd Currie tells how Ron Rokosz was part of a team that had to dissect each task at each level in specific Task-Condition-Standards (T-C-S) terminology. Early on, Rokosz's Company B was organized with a First Sergeant, Executive Officer (XO), supply sergeant and one clerk. This small element was charged with working on T-C-S's for all assigned missions in the Battalion with a great emphasis on meaningful training schedules. Both Rokosz and Currie remember spending many late nights with Rokosz's Executive Officer, Jerry Boykin (a future three-star special operations general) hand writing upcoming schedules.

Many members of the original cadre recall that in the Infantry Center there was a wall displayed with all the 3 × 5 cards with each T-C-S written and organized by level of command. The team would have long, excruciating murder boards where the wording was revised and reworked until the team thought they got it right. Montano was put in charge of collecting the finished product which would eventually evolve into a Light Infantry ARTEP (Ranger) which was provided to Forces Command (FORSCOM) and the Training and Doctrine Command (TRADOC) to exercise.[8]

The amount of time spent developing 3 × 5 cards with T-C-S proved beneficial. Every task from individual to battalion was defined. This mission took most of the year to complete. As the Battalion moved from Fort Benning to Fort Stewart the massive amount of T-C-S cards were taken off the wall of the Infantry Center and posted to the Operations wall at Fort Stewart.

PFC and later SSG Steve Hawk from B Company, 1st Ranger Battalion was a stellar young Ranger who was actually the honor graduate of his Ranger Class 6-75.[9] He recalls that there were long wooden 90 mm recoilless rifle boxes that carried the lot of the 3 × 5 cards with Tasks, Conditions and Standards. Hawk remembers that the leaders would randomly pull a card from the crates and then make their teams and squads train on them as a way of validating the T-C-S that was developed for that task. He said that once they got to Stewart, from July 20th forward, they would go through this exercise, not so much as training, but rather testing them to see if there's anything that had to be re-written or refined. This iterative process validates how meticulous Leuer was in defining strict performance standards.

CSM (retired) Mike Cheney looks back on the quality of training, discipline and leadership in the Battalion at that time with respect or Leuer's approach to the "lead by example" example. To this day, Cheney still has his 3 × 5 cards that contained the task, conditions and standards the Battalion developed that year. Later in his career, he used them in the Special Forces to train the Jamaican defense force. What was salient to Cheney that nothing was notional, everything was hands-on; 85% wasn't good enough.

Performance Oriented Training focus didn't stop after the initial volley of training was conducted by the Ranger Battalion. This iterative education process continued. The company commanders went to the Infantry Center for formal classes where they honed their mastery of this process and methodology in an academic environment. For example, they took a task as simple as "making a pot of coffee." The students would identify the given parts that included a pot, some volume of water, and coffee. They then would define the essential steps which included plugging the pot in, turning it on, etc. Therefore, if you forget to plug in the coffee pot you don't get any coffee and then the task is failed. This type of approach, once learned, can be easily translated to all a Ranger does.

On a side note, CSM (retired) Jimmy Broyles who was a Sergeant in 1974 and working for CSM Gentry recalls his coffee making talents didn't necessarily meet the standard set by Gentry. Broyles had met Gentry in Lima Company Rangers, 75th Infantry in Vietnam while Gentry was a First Sergeant. One night while assigned to 1st Ranger Battalion Broyles had the task of performing Noncommissioned Officer Staff Duties (SD), essentially it is a task for being the senior person awake in a unit at night to handle whatever may occur.

That night, before going home, Gentry reminded Broyles to make his coffee about 0400 so it would be ready when he came in for the day. Broyles told Gentry he didn't drink coffee and did not know how to make it. Gentry went into detail about how it was done, along with putting a small dash of salt into it to take the bitterness out.

At 0400 Broyles promptly made a full pot of coffee in that 'big ole machine' and calculated that a dash of salt for each cup – unfortunately his measure was off for he translated a dash for nearly the whole salt shaker. Gentry arrived at 0445 and while talking to Broyles he took a big swallow of his freshly brewed coffee from the big German beer mug he used as he walked with his back to Broyles into his office. Broyles saw Gentry shudder. He turned around and shaking the reports at Broyles and the staff duty officer and said, "Never, Never touch my coffee again!!!!!" Proving that concise measures in any task is critical for success.

By using clear, definable, measurable tasks, conditions and standards it was possible to further sub-divide these tasks into appropriate rank and teacher responsibilities; a level of specificity that Leuer wanted to identify in this process. For instance, the task "platoon in the night attack" would have clear requirements and standards for the platoon as an entity, each squad in the platoon and for each team in the squad.

Additionally there were tasks, conditions and standards drawn up for each level of leadership to include the platoon leader, platoon sergeant, squad leader and team leader as well as special duty positions like "radio operator" or "communications specialist." That meant each level had to teach its level for achievement of the standard. This was a KEY point. Everyone from top down was personally responsible for his level's success and it had to be clearly measurable in terms that all could understand. No one could skate. This was a main reason many mid-grade NCO's resigned from the battalion or were relieved. It was a mid-level void. These noncommissioned officers couldn't do the work but their subordinates could.

Performance Oriented Training put a clear definable burden on the noncommissioned officer corps. Noncommissioned officers were the heart of a unit and its success or failure was largely due to the energy, knowledge and tenacity of the NCO vice the officer. Leuer instinctively knew that if the Rangers were to be a success and to be an ultra-high performing organization, it would depend upon the buy-in and aggressiveness of the Noncommissioned officer corps-particularly at the squad leader level.

As they suited up for their jump and subsequent training in Fort Stewart, Leuer thought with pride of how he had built his cadre of Rangers at Fort Benning. He was mostly proud of the standards they built, the foundation they planted, and the rigorous thinking they did in devising a training plan to measure their capability and readiness for war. It was now time to train.

TRAINING AT FORT STEWART

On July 1, 1974, the Rangers loaded C-130 Air Force Aircraft at Lawson Army Airfield in Fort Benning, Georgia and conducted an inaugural parachute jump into Fort Stewart, Georgia. This act was the first official welcome for the Ranger Battalion to their new home. Thirteen days later, Leuer moved the Battalion to a place called TAC X, a large training area deep in the swamp and ultimately why the Rangers called Stewart "Camp Swampy." TAC X

1st Generation Airborne Rangers getting ready for a jump on Taylor's Creek Drop Zone in Fort Stewart, Georgia. (Source: Currie)

had been used during the Vietnam era as a helicopter training area. To support this training the US Army had built barracks-type buildings on location.

The airborne jump that day was onto Taylor's Creek Drop Zone. There were roughly 120 jumpers.[10] Ron Rokosz and Keith Nightingale were jumpmasters for each aircraft. As senior leaders, Leuer felt it appropriate they performed the duties as jumpmasters for this important jump. Leuer, like most of his Ranger parachutists, was a master rated jumper. Most of the officer and noncommissioned officers were also either master or senior rated. Ron Rokosz was the only commander who was novice rated at the time. Rokosz came to the battalion with five jumps; the requisite amount needed to graduate from the three week airborne school. Nowadays commanders who lead airborne units are typically senior or master parachutist rating when they are selected for command.[11] Rokosz would go on to be one of our nation's finest paratroopers leading the Ready Brigade of the 82nd Airborne Division on a no notice deployment to Saudi Arabia in the 1990-1992 Desert Shield and Desert Storm combat operation.[12]

Additionally there was about 20% of the Battalion on the ground at Stewart from the Battalion during the jump. After exiting the two C-130s the Rangers assembled and marched

Taylor's Creek Drop Zone in Fort Stewart, Georgia. (Source: Currie)

Ranger 90 mm Recoiless Rifle Section. (Source: Currie)

into the cantonment area to post headquarters for welcome by Col. Fred L. Dietrich, the Post Commander and a veteran of World War II airborne operations.

MSG (retired) Don Feeney recalls that on July 15, 1974, a few weeks after the jump into Fort Stewart, he was a new Sergeant or E-5 fresh out of the Noncommissioned Officer Candidate Course (NCO). Feeney was standing with 1LT Roger Brown, his platoon leader from 1st Platoon, B Company, who would go on to retire as a major and a member of the Ranger Hall of Fame. The two of them stood in front of Battalion Headquarters and watched as the cadre, who had been forming at Fort Benning, marched proudly down the street of Fort Stewart carrying the 1st Battalion, 75th Ranger guide-on. A photo of that event hangs on his wall to this day.

After the jump the Rangers began training in some form or fashion almost immediately. The initial days were immersed with heavy doses of physical training, drill and ceremonies, road marching, and live fire training. Every afternoon they would do the "Ranger Mile;" the term they used for a four mile tank trail run wearing load carrying equipment (LCE), helmet and boots. The course had six obstacles to negotiate along the way.

Leuer describes the initial training. "… Everyone is coming in here and we're training hard. We're running, we're shooting, and I'm implementing this performance oriented training, so I've got people doing task analysis, tearing a mission apart for its critical elements. They are developing the training for each of those elements and the standard that you have to meet in order to be successful at the top. I'm training cadre to do this; I'm learning to do it and were all working together. …"[13] The unwavering intention to train anytime was established early on in the initial training of this new force. To Leuer, few things trumped an opportunity to train.

Private Joe Hill enlisted for Ranger School following his brother's footsteps who had instilled in him the meaning of being a Ranger. Hill's brother was in K Company, 75th Rangers in South Vietnam and was later KIA while serving in a reconnaissance platoon in January 1971 while with the 173rd Airborne Brigade. While Hill was in jump school the Commander announced that anyone going to Ranger school or a follow-on assignment and had an interest in joining a new Ranger Battalion should talk to a representative that was conducting interviews. Hill went to the interview and it was there that he met 1SG William Skinner. Skinner asked him if he had gone to the Reserve Officer Training Corps (ROTC) program in college. Hill told him no that he just graduated from high school and joined the Army. He then asked him what he knew about Ranger School and Hill told him it was about endurance and that you couldn't be a quitter. Skinner then quizzed Hill what he knew about operation orders, leadership, patrolling techniques, small boat operations, knot tying, hand to hand combat, rappelling and the like. Hill told him he knew how to spell all those things and that was about it. Skinner said that he didn't think Hill was yet prepared to go to Ranger School, but if he came to the Battalion he could still go and when the Ranger Battalion sent him he would be better prepared. Hill accepted.

At age 18, fresh out of Airborne School, Hill headed to his first duty station with a good friend he had met in Basic Training. The pair left Fort Benning after jump school graduation and four hours later arrived at Fort Stewart. SSG Curtain met them at the Battalion when they arrived at about 2200 that night. Curtain gave Hill and his buddy their equipment and told them to put their web gear together, put one set of jungle fatigues on and the remaining five pair in their rucksacks. Everything else they put in Hill's car. The pair had just arrived at the Battalion and before they could sign in or empty the car they were mustered out to the field to train. Hill and his friend got in the battalion jeep with Curtain and headed out dressed in new jungle fatigues, new jungle boots, helmet and rucksack.

They showed up in the dark somewhere and Hill met 1SG Skinner for the second time. He welcomed them to A Company, 1st Battalion, and then told them that they must be in pretty good shape since they just graduated from jump school. They assured him they were. Skinner then announced that they were going on a 25 mile road march. "Welcome to the 1st Battalion", he said. It must have been midnight when they began their trek. Five hours later Hill was introduced to Weapons Platoon and shown to the barracks. Hill's comment on this experience is one filled with pride when he exclaimed, "Only in the 1/75th Ranger Battalion are you in the field before you're in the barracks."

While the success of the initial train up and the subsequent jump into Fort Stewart made it seem like the path was an easy trail, it was quite the opposite indeed for the timeline and key goals and objectives before moving to Stewart were never quite set. In fact, Leuer didn't know they were leaving Fort Benning for Fort Stewart until sometime in April 1974. But like many a subordinate leader said of Leuer, "he knew the goal, how to get there, then took us there."

Benning was the initial choice to house the new Ranger Battalion. When the decision was made to move to Fort Stewart buildings were erected to support the new unit on their new post. Regardless where they were housed, little slowed the Rangers down in their intent to train.

While they trained they didn't stop doing PT. The initial run standard was five miles in 40 minutes; or eight minute miles. When the bulk of junior enlisted Rangers arrived at Fort Stewart the standard dropped to five miles in 45 minutes; however that standard remained "that slow" for just a short period of time.

All runs were done in boots. Road March standards set were five miles in one hour – in full gear to include mortars and recoilless rifles. These marches progressed to eight miles in two hours, 12 miles in three hours, 16 miles in four hours and finally, 20 miles in six hours with the distances and times first validated in December 1974. Don Bruce would say later that these weren't road marches, but rather sprints with full TOE equipment!

The author recalls 25 years after the standup personally doing road marches at these distances and times. Doing 20 mile road marches with 35 pound packs monthly and 25 miles with 55 pound packs quarterly doesn't of years later proved that the standard stuck with the Rangers as a prerequisite to continuous service with this elite force.

A month after they jumped into Fort Stewart the Battalion claimed full strength by August 1st. For accountability purposes the Rangers were authorized ten percent over to account for those non-available due to leave, injury or schooling to ensure the Battalion always had 100% on hand. Leuer was fortunate to have the vast majority of his Rangers in place when the training began in earnest.

When they trained they trained hard, very hard; it was the epitome of intensified training. Everything was a live fire of some sort. Hawk recalls that they had little time off and worked on weekends, taking Sundays to prepare for the next week's training. When the rest of the Army was cutting back on training and training ammunition, the Rangers got more and more resources especially life ammunition. According to Leuer he didn't think they fired fifty blank rounds in the six months.[14] Training was tough and demanding but no one got hurt. Leuer's motto was that – you don't train to hurt people but if you don't train and train right you will end up hurting people other than the enemy.

September to December 1974 was a critical transition time for the testing and evaluation for validation of the Battalion as a combat ready force. Fort Stewart was a sleepy little post with only a few support units at the time of transition. The decision was made to activate the 24th Mechanized Infantry Division on Stewart, but that did not begin to occur for several months. Fort Stewart offered extensive training areas for both maneuver and live fire operations. The post encompassed 280,000 acres, stretching 34 miles from east to west, and 19 miles from north to south.[15]

Given the light structure of the Rangers, Leuer attained the operational support for the Battalion from Fort Stewart units like the 92nd Engineer Battalion, the 24th Ordnance Company and the 84th Transportation Company. The engineers built the Rangers a 34-foot jump tower as well as hand-to-hand pits to allow the Rangers to practice their trade. Gaining this type of strategic support from units that shared the same posts as the Rangers would be the norm until a special operations support structure was put in place some years later.

The long awaited and anticipated training commenced as junior enlisted soldiers continued to arrive at Fort Stewart. The performance oriented standards continued to be defined and redefined throughout the period. In fact, From July to December 1974, Ron Rokosz recalls an extensive training period where the Rangers at Fort Stewart executed Expert Infantry Badge (EIB) training and testing, squad emergency deployment readiness exercises (EDREs), extensive live fire operations, and a progressive series of road march standards to the final standard of twenty miles in six hours with a complete set of TOE with five miles in one hour as the toughest standard. The Ranger commands also conducted extensive airborne operations, almost all at night, dropping from as low as 600 feet to stimulate combat jump conditions.

Leuer confirms that the training was intense in this short period cramming squad, platoon and company level training at the Fort Stewart TAC X training area. Each level had specific measured evaluations to ensure his Rangers were meeting the performance oriented standards. If a standard wasn't met, the Rangers were made to train on it and exercise that task over and over again until the measure of success was met.

Mike Wagers said that the time spent at TAC X, especially in the final ARTEP in December 1974 was more than a challenge. In fact, he flavors the ARTEP as the "most horrendous field exercise ever experienced. Twelve days in duration, culminating in a 20 mile road march." Wagers estimates that his platoon moved over 100 kilometers prior to the 20 mile road march.

All in all it was a seminal and memorable time for the new Rangers. Ron Rokosz remembers going on a reconnaissance of Fort Stewart with A Company Commander Don Clark. He recalls flying over training area and seeing a crocodile in the water and knew right then things would be different. The terrain at Fort Stewart was varied with many swampy areas. Rokosz recalls negotiating land navigation courses which challenged his Rangers as they attempted to navigate in chest deep swamps.

To Rokosz, his command strove to make training as realistic as possible. They practiced extensive use of live fires and developed different means to evaluate effectiveness of fires. He recalled Gen. DePuy visiting his company and walking squad live fire lanes with him. This gave the TRADOC Commander first-hand knowledge of how "his" performance oriented training was coming along and he was pleased with what he saw in Rokosz's Rangers. Jerry Barnhill recalls emphasis on correctly teaching the DePuy fighting position which provided for interlocking fires and then having Gen. DePuy visit his company and being impressed with how the Rangers embraced his concepts.

Frank Magaña of Rokosz's B Company, and the SSG Acebes disciple, recalls DePuy visiting Fort Stewart and coming to his squad's live fire exercise. During this particular exercise, DePuy took over as squad leader and was maneuvering and bounding the squad through the live fire exercise.

As a junior enlisted man Magaña was impressed with the detail and level of training in the 1st Ranger Battalion. He recalls a Pathfinder field training exercise (FTX) at TAC X where private first classes and specialist were taught how to set up and mark landing zones (LZ) and drop zones (DZ), how to rig helicopters for rappelling operations, and how to conduct resupply drops. Most of these were tasks for noncommissioned officers in a typical army unit but the Rangers were different; if they wanted to "grow their own" leaders they had to start early and be relentless in the training of these potential junior leaders.

Additionally, Magaña thought quality of emphasis and precision on Drill and Ceremonies (D&C) was instrumental in building unit discipline and cohesion. For D&C the Rangers marched at right shoulder arms with fixed bayonets, while the rest of the Army marched at sling arms. This was another way for Leuer to instill discipline; doing simple things perfectly, by the numbers so units could do the manual of arms while marching. For example, if your elbow was supposed to be at 90 degrees at right shoulder arms, it's 90, not 85 or 95. Leuer said if he wanted to assess the combat readiness of a unit, he would have them assemble on a drill field and do drill. That would tell them how disciplined the unit was.

Rokosz was a no-nonsense commander. The author can attest to this some 16 years later as he was the Headquarters and Headquarters Commander for Rokosz when he commanded the 325th Airborne Infantry Regiment (AIR) in their no notice deployment to Desert Shield in August 1990. One of the things Rokosz's soldiers remembered from his time as a company commander of B Company was that he loved gassing his soldiers. Every chance he got he would toss a CS (tear gas) grenade into the field exercise.[16] He was determined to ensure his Rangers could execute the combat drill of donning their protective masks; an effort that proved fruitful in the early days of Desert Storm when his paratroopers were susceptible to a possible Iraqi Scud chemical attack.

Whatever the case, Rokosz was determined to focus on tough, realistic training.

During this period B Company was given the task to assess the realism of the go-to-war basic load. Rokosz deployed his command to TAC X with each soldier loaded to bear. They were weighted and photographed and then stolen off on a twenty mile road march. They were tasked to determine if the defined basic load for a Ranger company could be properly man-packed and carried in combat. When the company arrived at TAC X the Ordnance Company brought out an entire basic load with 60mm and 90mm rounds, small arms ammunition, grenades, flares and other sorts of ammunition and supplies that would be required by the Rangers to survive a three day mission. B Company conducted a 20 mile road march with that load.

Of course in good Ranger fashion it rained the day before and the tank trails were very muddy. This "test" became a grueling and sloppy march. The Rangers learned a lot about themselves that day. One mortarman who weighed a mere 126 pounds was weighted down with 130 pounds of load. It was clear that the Rangers had to practice and train more under these types of loads and requirements. Leuer directed the Battalion Logistics officer or S4 to obtain dummy ammunition loads which they used later in training to simulate combat weight.

According to Nightingale, everyone practiced the load bearing challenge. In his headquarters company the ability for his soldiers to carry their mission essential equipment "became almost unbearable" for the medical and signal soldiers, for instance, had an abundance of special items that had to make it to the battlefield. The only way to get it there was on their backs. When adding the weight of ammunition with task-specific gear, and then food and water and their personal gear it was nearly impossible for these "low density soldiers to make a 20 mile road march without falling out.

Mike Wagers recalls that at some point in this period Rangers learned what how much everything weighed. We knew what the combat load weighed right down to each member of a squad. The Rangers eventually knew how much a weapon, helmet, LBE, and rucksack weighed, right down to the equipment required of each duty position. Machine gunners knew exactly how much the load was for the assistant machine gunner. Wagers recalls that the basic load for the rifle platoon sergeant was the lightest load in a rifle platoon, weighing in at 109 pounds.

Included in this intensive training time was specialized training that included boat operations, rope bridging, rappelling, hand-to-hand combat, and pathfinder operations. Leuer ensured that all mission essential tasks were exercised and performed to proficiency – all with the intent on achieving a "go" on the battalion ARTEP in December.

Leuer talks with much excitement and pride when it comes to describing the training period. The squad live fires were realistic with overhead fires coming from machine guns as the squad leader maneuvered his squad. The Rangers came up with creative ways to conduct range firing like putting balloons on silhouette targets to evaluate marksmanship. When Gen. DePuy visited and saw the balloons he joked that the Rangers were state of the art, and commented about how far behind the Army was in realism and evaluation standards.

The use of aggressors in force-on-force training came to the forefront during this period. The Battalion intelligence officer, or S2, was responsible for the aggressor part of the training ensuring the Rangers got the appropriate "foe" to fight against. Leuer recalls training the squads during night attack operations that the squads would deploy in the prone and then be required to low crawl forward of the line of attack. The S2 ensured an evaluator sat past the line on a metal chair for a while and if he detected the squad in their movement he would blow his whistle. This would go on throughout the night until the squad made it past the line undetected.

The Rangers tried to take advantage of every training opportunity possible. They often went back across the state of Georgia to Fort Benning to shoot and train. The command for Fort Benning at the time was Maj. Gen. Thomas Tarpley who had been Leuer's division commander in the 101st Airborne Division in Vietnam. Tarpley wasn't much of an "out of the building" or "out of the bunker" type of leader, regardless he embraced Leuer's decision to deploy the entire Battalion back to Benning to shoot in September 1974. Tarpley had great confidence in Leuer and was able to give him good advice because of their shared experiences from the 101st.

The training culminated for this newly formed unit with the ARTEP evaluation in December 1974. The ARTEP was administered at Fort Stewart and ran for 12 days consisting of discrete events, where a squad or platoon would be pulled out and evaluated on their ability to complete a designated mission. LeMauk recalls that they planned a number of sequential missions requiring the Rangers to move quickly from one objective to another with little time available for planning for rehearsal. Rokosz recalls that it was cold and nasty but the Rangers had to do well for the future of the Rangers depended on the outcome.

In the Rangers, you kept going for the full time – no one even considered giving less than 100%. This attitude, which extended well beyond physical activity, provided a superb foundation for tackling every aspect of military life. You couldn't always be the best, but you could always try – something most found lacking in many other units. The ARTEP that December culminated with a 20 mile road march back into Fort Stewart, all within a six hour standard.

1SG (retired) Steve Rondeau was a specialist at the time and recalls this exercise was designed to show the Department of the Army (DA) that the 1st Ranger Battalion was combat capable. Rondeau was in 1st Squad, 1st Platoon of A Company and his platoon evaluator for part of the field training exercise (FTX) was MSG Gil Berg. Many remember him as "General Greb," a fictional target of many Ranger missions and a formidable foe. However, in this first meeting, MSG Berg was in awe of the loads the Rangers carried and the swamps they waded.

One mission called for the company to move to a pickup zone (PZ) and airmobile to another location. Rondeau and his Rangers waded through cold dark swamps for hours. He remembers still being in the swamps and hearing the first lifts starting to go. They picked up the pace under threat of being left and made it out to the PZ. Rondeau and his Rangers did not do a combat cross load and most of 1st Squad was on one UH-1H (a utility cargo aircraft).

Not long into the flight a loud bang and grinding sound came from the engine area. The aircraft lurched and the crew chief started yelling to jettison rucks. The squad leader, SSG Reynolds was directing the Rangers to maintain their rucks and almost immediately the Rangers were told the aircraft was going to crash.

The aircraft had suffered a tail rotor failure. The emergency procedure for a UH-1H tail rotor failure was to make a "running landing". The aircraft would fly reasonably straight if the speed was kept over 90 knots. Slower than that, the engine torque would cause the body of the aircraft to rotate. Fort Stewart had many large, wide, sandy roads and the pilots did a good job of picking one out and making a perfect running landing.

This happened so fast that there was a lot of confusion and yelling followed by a lot of sand as the aircraft slid down the sand trail. The squad followed procedures and assembled off the nose of the aircraft. The pilot and co-pilot were sitting on the ditch bank sharing a cigarette with shaky hands while the crew chief stood with the tiny, on-board fire extinguisher looking at the bird.

No one knew exactly where they were. The pilot and Reynolds agreed on the location of the landing zone (LZ) and that the company knew their situation and would wait at the LZ until they could link up. The pilot had used the radio to relay though another UH-1H that had orbited overhead to render assistance in case the landing went bad.

After an equipment check the squad departed, road marching; still wet and cold. It seemed like the Rangers had traveled 10 miles before they found and linked up with the company. After linking up the Rangers moved the rest of the night with dawn breaking before they stopped.

Another ARTEP story that Rondeau recalls is about the battalion moving into a defensive trace along the Fort Stewart impact area observation and firing points overlooking the Canoochee River. Here the new DePuy fighting positions were dug and live ammunition was passed out – and kept getting passed out. Every weapon had plenty of live bullets. Mortars, recoilless rifles, grenade launchers were all loaded up. This was the Battalion's basic load of live ammunition. Rondeau wound up with two cases of match ammunition for his XM-21 sniper rifle. The range was "hot" and fire commands would come without warning.

It was good for the leaders to practice their command and control and fire discipline. The lack of targets caused some interesting fire commands like "M-60, front, pine tree in the open, range 100 meters, fire." As Rondeau recalls the machine gunners in his sector had fun using their Traversing and Elevation (T & E) mechanisms to trim the limbs off a selected pine tree and then cut the tree down.

It was raining and the fox holes were over knee deep with rainwater. To stay out of the water and mud filled foxholes, the Rangers lay beside the hole and fired. Word came down the line that the Fort Stewart post commander, Col. Dietrich, was walking the line. The area got a quick policing, camouflage touched up, and the Rangers were in their holes standing in the water when he came up with a small crowd of evaluators and our own folks. He looked at Rondeau and said "Son, do you have dry socks on?" Rondeau was standing knee deep in water and just a little dumbfounded by his question. He recovered and said, "No, Sir, but I do have plenty of dry socks and foot powder in my rucksack." He said that was good and went into a short speech about infantry soldiers and the importance of healthy feet.

He began to walk off and stopped. He turned and looked at Rondeau and frowned. He looked at the positions to his left and right. Then he said "Son, why are you armed with an M-14?" The crowd around him got big eyed and started digging for their note pads. I'm sure they thought some kind of major screw up had been uncovered. Rondeau said "This here is the XM-21 sniper system. I am a school qualified sniper." He had several more questions about the weapon and his scope and how far he could engage targets. When he handed the weapon back to Rondeau, he looked out over the impact area and inquired if it was possible for him to shoot something for him.

Rondeau found a mortar flare hanging in a tree about 400m away and told him he was going to hit the metal base. The crowd started sharing binoculars and murmuring that it was too far. Rondeau was getting ready to take the shot when a blur of white went through his sight picture. He raised his head to see a white, crane like bird, settle into the swamp about 700 meters out. He told the Dietrich he would rather shoot the bird if that was OK with him. The staff guys behind were trying to warn Rondeau and were shaking their heads "no."

Rondeau took the shot and was able to get back on target in time to see water and mud splash behind it. The bird never moved. He heard "He missed" come from the evaluators. About that time the bird realized he was dead and fell over. Dietrich left the area after being assured that all the Battalion snipers could shoot that well. Rondeau's company commander returned and chewed him out for "show boating" instead of taking an easier shot

In December 1974, Jerry Olsen was a PFC and a member of B Company. He remembered that the Battalion ARTEP was a three-week field exercises for Company sized operations. Olsen was serving as an ammo bearer as part of a 90 mm gun team in the B Company Weapons platoon. Weapons platoon was split up to serve with the line companies for this exercise and SSG Peter Lynch, SPC John Nowak, PFC Bill Winget and Olsen were assigned to Hog Brown's 1st Platoon.

About halfway through the exercise, they were on a company sized mission which they trudged through some swamps and thick underbrush during the day in a line formation. The gun team was in formation at the back of 1st Platoon with SSG Lynch, Winget, Nowak and finally Olsen, bringing up the rear. 2nd Platoon was following Olsen in formation, generally 3-5 meters apart even through the thick swamp. While following in line, Olsen noticed that

SPC John Nowak and the B Company Weapons Platoon on an ARTEP in
Europe in December 1974. (Source: Currie)

A Ranger rappelling in
Panama. (Source: Currie)

Lynch had stumbled a little bit. Next Bill Winget did a sort of left flanking movement and was flailing his arms and getting stuck in the brush, though he was maintaining good noise discipline. Olsen was still oblivious to what was happening but thought it was kind of funny that Winget was stuck in the brush because of the 90 mm recoilless rifle he was carrying.

Just then, a black cloud of yellow jackets swarmed over Lynch, Winget, Nowak and Olsen. It now became abundantly clear to him what was happening. Lynch had kicked a yellow jacket hive and the residents of that damaged hive were getting their revenge on their poor souls. The yellow jackets were everywhere and Olsen looked behind to see that 2nd platoon had already identified the issue and had reacted quickly, diverting their movement away from the kill zone and veering off to the right.

The yellow jackets were everywhere stinging them all over their bodies. There was no escape. The group had to keep walking in formation and suffer while simultaneously attempting to kill as many of these attacking insects as possible by swatting at them or crushing them when they landed. Olsen recalls that he was in total agony and feeling sorry for himself when he saw John Nowak covered in yellow jackets including several on his face and one in particular affixed to his lower lip. He smiled at Olsen, apparently enjoying his misery. That smile forced Olsen to see the twisted humor in all of this.

When they finally hit dry land they took off their rucksacks and web gear. Olsen opened and un-tucked his t shirt and a bunch of dead yellow jackets fell to the ground and a couple flew off. Most of them got stung multiple times that day with around 50 or so stingers each. The initial assault wasn't the worst thing however. They had to suffer with the constant pain and irritation of the stingers for at least the next 10 days or so while in the field. Olsen describes this as just another fine day to excel in the Ranger battalion.

All in all, the ARTEP was serious stuff and the Rangers worked very hard to perform the standard. Gen. Warner visited the Rangers during this period to gain an assessment for how they were progressing. Warner visited the Rangers at Fort Stewart no less than four times in 1974. He was very impressed with the discipline and quality of the soldiers.

His latest visit and report had a big impact on the rest of the Army. He recognized and reported that the Rangers weren't just some unit stealing people, but rather a capable combat ready force. According to Leuer, after the Warner visit, the rest of the army started picking up on them. At first most of the Army didn't think the Rangers could instill such discipline and train to such a standard, given the state of the army at that time. However, as Leuer puts it, the light got lit and other people started looking around and soon the rest of the Army became believers in what his Rangers could accomplish.

Walking the ARTEP lanes with the Rangers, Warner recognized right off that it was a strenuous military exercise especially the road march coming back from TAC X. To Warner – and Leuer – this was a litmus test. Any unit that could endure the evaluation as strenuous as a measured ARTEP as undertaken in the swamps of Georgia and the intense road march they conducted to standard was a formidable force. Warner quickly made a few phone calls and then called Leuer in to tell him that he just reported to the Department of the Army (DA) and Forces Command that the Rangers were a combat ready unit and never in his 28 years has he been with a finer group of soldiers and officers that Leuer had in his organization. He told Leuer that he can be proud of what he had done in so little of time.

This evaluation process was initiated by Leuer allowing him to give a critical eye to this performance based evaluation process – which ultimately helped in developing and codifying the performance oriented training concept which led to a universal standards based lifestyle for the Rangers. This long term training and evaluation process was exercised in draft at Fort Stewart by a team of outside evaluators from Stewart and Fort Bragg. The training eventually migrated to Fort Bragg where the first real ARTEP evaluation was conduction in 1975 by the 82nd Airborne Division. Wagers recalls that during the evaluation the 82nd evaluators had to be rotated as they could not keep up with the pace of the Battalion.

ARTEP evaluations were eventually retired from the Army lexicon and replaced by very thorough Mission Rehearsal Exercises (MRX) that are a validation exercise US Army units undertake, mostly at the Brigade Combat Team (BCT) level as they prepare for a deployment and integration into the Force Generation Cycle.

With the conclusion of the exercise at TAC X in December 1974, Phase III of 1st Battalion, 75th Rangers' activation began. Phase III consisted of worldwide specialized large scale training in various climate and terrain conditions. These exercises were all over the world and conducted in all environments to include amphibious, arctic, and desert.

THE ARTEP CULMINATES WITH BANNER DAY

At the end of the initial train up at TAC X Leuer called for what he termed "Banner Day." As a custom since that period, once a month the Battalion would, as Leuer put it, "pull in, clean everything up, and have a day of athletics."

These days were a down day when the subordinate commands in the Ranger Battalion could have some camaraderie, competition on the athletic field and espirit-de-corps. For this event, the Rangers made a banner that was consisted of the six colors of Merrill's six combat teams.[17] Whoever won Banner Day got to don the Merrill's colored banner on their guidon for a month as the Battalion winners until the next Banner Day.

Frank Magaña recalls the first Banner Day held at TAC X. B Company was leading by only a few points as the battalion went into the final event which was boxing. SSG Acebes pulled Magaña aside and told him he was going to box for the company. Magaña did not question it because he felt that if his squad leader had confidence in him then he had confidence in himself. He won his bout, and as a result, B Company won the first Banner Day.

Rokosz thought that Banner Day could give his company and his Rangers a competitive edge. Leuer habitually injected these events at the end of training periods. These events, as described by Keith Nightingale, were the a full day of athletic competition between companies – all very physical and demanding – events like pushball, combat soccer, rope pull, relay races and other physically competitive events and culminating with boxing events. Rokosz's B Company won the first Banner Day in total. Nightingale's Headquarters Company, at only 50 strong, had to field teams for all events so Banner Day was particularly grueling for them. Nonetheless, Leuer ensured that every Ranger regardless of rank participated.

This competition was a big deal to the Rangers. Regardless of the size of the unit, and Headquarters and Headquarters Company was at least two thirds smaller than rifle companies, everyman in the company had to play. Leuer and the senior staff were all part of each team. Nightingale recalls that Push Ball was the big killer for it was physically demanding and messed up lots of his folks in the competition. Headquarters and Headquarters Company didn't have the comfort of specializing its "players" since they were so short everyone had to play almost every event. As Nightingale recalls, "the big issue was that we didn't get any breaks due to numbers." Everyone had to play all the time and it was a balancing act to get people to simultaneous events.

Ron Rokosz on
Banner Day.
(Source: Currie)

LeMauk recalls that Banner Day was a field day for our medics, who practiced their skills as if on a battlefield, applying first aid and sending Rangers back into the game with stitches or a makeshift cast. The only thing missing was bullet holes. And no Ranger dared remove himself from a game – the only way to leave was unconscious. In retrospect, it was probably a little rough. At the time, it was how the Rangers did business.

Celebrations like Banner Day gave the Rangers an opportunity to recognize their successes achieved in such a short period of time. Once the initial train up of 1974 was completed, and the Battalion was validated as combat ready, they began to focus on specialized training. Just under nine months after the Ranger Battalion received its orders to activate, the 1st Ranger Battalion (Infantry) was officially activated on August 20, 1974 at Fort Stewart, Georgia. This date is honored every year as the official anniversary. That day the Secretary of the Army Howard "Bo" Callaway and Forces Command Commander, Gen. Walter T. Kerwin reviewed the troops along with Leuer and Colonel Dietrich.

Secretary Callaway passed the Battalion colors to Leuer, and he, in turn, passed the company guidons to the company commanders. Leuer believes that this may have been the only time in US Army history that a Secretary of the Army participated and passed the colors in a battalion activation ceremony.

After the activation ceremony Callaway wrote a letter to Abrams on August 22, 1974, who was in the hospital at that time and would pass within a month of receiving Callaway's letter. Secretary Callaway wrote:

> I participated in the official activation ceremony for 1st Battalion (Ranger), 75th Infantry, at Fort Stewart on Tuesday of this week. I just wanted to tell you that, in

Secretary Callaway passed the Battalion colors to Leuer, and he, in turn, passed the company guidons to the company commanders. (Source: Currie)

my judgment, the Rangers are everything that you had hoped they would be. I've never seen a unit that looked better or one which had a higher sense of mission, profession, and pride than this battalion of Rangers. I had an opportunity to talk individually to a great many of the soldiers. Every one of them believes that the Ranger Battalion is the greatest thing that's ever happened in the Army.

None us know what missions the Battalion will be called upon to perform, but I can promise you that anyone who choses [sic] to tangle with them should be prepared to "bring their lunch".[18]

Callaway was clearly impressed with the Rangers. The caliber of officers, noncommissioned officers and soldiers, all proudly referred to as Rangers, allowed the formation of the unit in an incredibly short period of time.

As 1974 came to a close the US Army was looking to rid itself of unnecessary footprints and force structure cuts were inevitable. Fort Stewart housed both the 24th Infantry Division and the 1st Ranger Battalion at the time. The commander of the 24th Infantry personally told Leuer that he should have little concern of the fate of his newly formed 1st Ranger Battalion (Infantry) for the entire Army at this point knew they were easily deployable and utterly reliable.

As Lt. Gen. (retired) Gary Speer, a platoon leader at the time of the activation of the 1st Ranger Battalion, states so eloquently, "the formation of the 1st Ranger Battalion set the course for the transformation of the US Army from Vietnam to the force without peer that fought in Desert Storm."

5

Rangers in Action – the Post Vietnam Era Operations

My troops are good and well-disciplined, and the most important thing of all is that I have thoroughly habituated them to perform everything that they are required to execute. You will do something more easily, to a higher standard, and more bravely when you know that you will do it well.

Frederick the Great, *Principes Genereraux,* 1748

In war, things get blown up really fast

Overheard from a US Army Ranger Sergeant

FORGING A FORCE

Once the Battalion passed their ARTEP in December 1974 the train started moving fast. Unlike in most post conflict eras, this time it looked as if the Rangers may be here to stay.

Every conflict since the formation of the 1st Rangers has had some Ranger involvement in some fashion. The 1974 Ranger force methodically formed into a larger force in the following years as it became more involved with the special operations community. Many missions were classified and remain that way. What is knows and included in this work is that there are a number of publicly known operations involving a significant inclusion of the US Army Rangers.

This chapter describes the missions requiring the employment of the US Army Rangers in the litany of conflicts following Vietnam. The Rangers were involved in every conflict since that time to include today's ongoing wars in the Middle East. This part of the book provides the reader a lively description of combat operations including Ranger missions; how they were employed and a review of the ultimate outcome. The wars in both Iraq and Afghanistan cover over a decade of deployments and this work explores the Rangers involvement in the beginning of each conflict and at salient points throughout each war. This gives the reader an understanding of the Ranger's evolution since their inception in 1974 and over time.

With the inclusion of the 1st Ranger Battalion into the Army inventory, the senior leadership was excited about what they saw in Leuer's ability to forge an elite force. When Vietnam came to an end both A and B Companies of the standing 75th Infantry remained on duty which helped field men and equipment to both the 1st and 2nd Ranger Battalions as they formed. Additionally, the personnel and equipment of B Company, 75th Infantry of the 5th Infantry Division (Mechanized) were transferred and the company deactivated on November 1, 1974.[1] To round out the Ranger force the 3rd Battalion was organized in 1984.

All three battalions were then placed under the command of a new Ranger headquarters, the 75th Infantry (Ranger) Regiment which was established in 1986.

The 1st and 2nd Rangers eventually found their homes and respectively placed their headquarters at Hunter Army Airfield (HAAF), Georgia, and Fort Lewis, Washington. The 1st Battalion and now the 2nd Battalion along with it continued to train and imbed with the Army's worldwide training exercise program. In the years leading up to and including the wars in the Middle East in the 21st Century the Rangers will have expanded their organization even more. Later chapters will expand on each of these subsequent changes.

The modern day Rangers were forged in 1974 and in the same year validated their abilities to be a formidable force. Their involvement with the existing exercise program solidified for the nation their abilities as an elite force. It also confirmed their readiness posture as the force of choice when the nation called for rapidly deployable combat forces.

CANADIAN EXCHANGE EXERCISE

In late to mid-summer, 1975 the Rangers went on their first Return of Forces to Germany (REFORGER) exercise.[2] This gave them opportunity to work a potential real world mission in a training environment with other forces in West Germany. Germany was the potential battlefield where the fight would occur with Warsaw Pact nations in the chance of an eastern invasion. This exercise, like others that the Rangers partook proved fruitful for the new command.

For some months prior to REFORGER there had been a competition within the Battalion between the rifle platoons. The competition was to measure which platoon did the best on the EIB testing, squad qualification testing and on a Platoon Tactical test. As Dave LeMauk remembers, the reason for the competition was to choose who would represent the Rangers in a platoon exchange with the Canadian Airborne Regiment. For a one month period the Canadian Airborne Regiment, (more specifically a platoon from D Company 2 Commando) would send a platoon to Georgia and in exchange 1st Battalion would send a platoon to Canada.

All members of the fortunate platoon from the 1st Battalion selected would be able to earn Canadian Jump Wings.[3] As with most competitions in the Rangers, the platoon sergeants were working their respective men hard. According to LeMauk zero defects as a measure of success was an understatement! Anything less than 100% on anything wasn't just frowned upon, it wasn't allowed. Everybody wanted to go. Getting the opportunity to earn foreign jump wings and, most importantly, the chance for adventure, was very important to the Rangers.

It did not take long for the field to narrow. It was finally down to three platoons; 2nd and 3rd from Charlie Company and a platoon from Bravo Company. After a stiff competition 3rd Platoon, Charlie Company under the leadership of 1LT Buck Garrison and SFC Mike Wagers got the nod to go to Canada for the competition. When the 3rd Platoon was selected to go to Canada almost every Ranger in the platoon of course was thrilled. The lone exception was SSG Herbie Baugh. His ankle was in a cast and he became part of the rear detachment; meaning he couldn't go and was surely disappointed.

Instead of another weeks-long field training exercise slogging around waist deep swamps in Georgia, the lucky Rangers were on their way to a foreign country for the first time as a unit. Most of the younger Rangers had never left the United States, so going to Canada as the best platoon in the 1st Ranger Battalion filled everyone with excitement and pride. The thought of earning foreign parachutist wings was a great incentive for everyone. Like most soldiers, the younger members of the platoon didn't always understand the bigger picture in the exercises, but it was a great opportunity working with the Canadian paratroopers and learning about the differences between the two armies.

Additionally, as is always the case on unit exchanges, they were eager to swap rations. In this case the Rangers saw themselves as big winners for the Canadian rations came in a box per day, with real jam, butter, and fresh crackers, and a wide assortment of brand name foods that they had actually recognized by name. When the Canadians tried the US rations, which tended to taste bland they quickly quit offering to swap. On the cultural side, the Canadian paratroopers offered to take the US Rangers downtown on nights off. They routinely ordered two beers at a time, alleging the waitresses were extremely slow. It didn't take long to realize the Canadians liked to drink fast in order to get their full before making it back to the barracks before curfew.

The actual training with the Canadians was a great experience. LeMauk remembers the mission overall was a successful operation and thinks the platoon "did the Battalion proud." Wagers recalls that the Canadian trip as it related to daily regimen in the Battalion. During pre-jump training, the US Rangers were shocked to learn that the Canadians didn't use a safety wire to ensure the static line snap hook stayed hooked. The Canadians may have been even more shocked to learn that the US parachutes weren't equipped with anti-inversion nets they had already used for years, something the US later added with great impact.[4] One of the most unusual experiences for the Rangers was jumping from an Otter, a small plane with room for five to sit on the floor, crabwalk to the door and spring out from a sitting position.[5]

LeMauk was on this particular mission and describes the jump out west onto the Island in British Columbia as "very exciting." He was in the back of the aircraft near the jump door, and could see out as the jumpmaster was making his checks. At the one minute warning, the aircraft was over water and stayed that way until just before the green light came on indicating it was safe to leave the aircraft. LeMauk then saw from the troop door the tallest trees he had ever seen, followed by thoughts that maybe they were in the wrong place. However when the green light came on, out went the Rangers, and below them was a beautiful, green pasture that promised a soft landing and an easy walk to the assembly area.

On the descent LeMauk noted that, all of a sudden, the Rangers could see the pasture was covered with cows and fences. Making it worse was the fact that the fences were barbed wire. A couple of Rangers landed on them, sustaining ripped uniforms and minor cuts, but all made it to the assembly area safely. The only thing on the island besides the objective set up by the Canadian military was the Army base and a dairy farm. On the aerial photos the drop zone (DZ) the Rangers didn't see all of the cross fencing and below that twelve inches of flowing green grass was about an inch of cow manure and several inches of water. Needless to say the jump, or more specifically the landing, was pretty messy.

The rally point was a dairy barn where two farmers stood laughing at the "Yank" Rangers covered in wet runny cow manure. The farmers had a big water hose that they used to wash out the dairy barn and LeMauk asked them if he could use it "to wash the boys off."

Jack Rogers (a retired Lieutenant Colonel and is well known as one of the last "originals" to leave the Battalion) recalls that once they parachuted onto the island they linked up with the Canadian Navy. Upon linkup they boarded a naval destroyer anchored in the harbor. After boarding, the destroyer immediately departed for the objective. The Rangers enjoyed this adventure making their way to the bowels of the ship for a warm meal and a briefing on the follow on mission. Just after midnight the Rangers climbed over the side of the ship into a landing craft and headed to shore for the attack as part of the exercise.

After landing the Rangers linked up with partisans who took them part of the way to the objective and then they disappeared leaving the Rangers to their own fate. Rogers just returned from Ranger school before this operation and was about 15 pounds underweight, but recalls that these were exciting times for a young Specialist Fourth Class.

One of the strangest things to happen to the Rangers on this particular patrol was when they surprised a homesteader chopping wood at night. Rogers said that the look on his face was priceless. He hollered, "Indians!" and took off at a pretty good clip down the side of the hill. Expecting the wide-eyed woodsman to return with other suspecting islanders they quickly moved out of the area. After some distance they finally arrived at the Objective Rallying Point (ORP) and Lt. Buck Garrison disappeared into the darkness with a small element of Rangers to conduct their recon of the objective.

Unfortunately, the incident with the traumatized axe man made Rogers lose his bearings. He was totally turned around as to his whereabouts which was frustrating for he had been drilled and drilled to always know your position down to an eight digit grid coordinate.

Rogers whispered to PFC John Mashburn, whom he called "the Preacher," to keep his eyes open while he did a map check. At this point it's about 0400 and they had been going non-stop for about 36 hours. Rogers pulled his poncho and map out and got underneath the poncho so that no light could be seen. But here lies the problem; his flashlight would not work. Rangers always joked about how a flashlight was a cylindrical object that held dead batteries. However, he did have a very small storm candle in his cargo pocket which he lit to use for light. He remembers staring at the contour lines of the map and thinking what a pretty pattern. The next thing Rogers remember is Mashburn whispering really loud about how he was "ON FAAR!!"

Apparently the warmth from that little candle was just enough to put Rogers into a "Ranger Trance" and he fell asleep. A couple of unfortunate events transpired; the poncho he was holding above his head burst into flames and the sticky burning goo from the melted plastic poncho acted like napalm and stuck to his jungle fatigues which immediately burst into flames.

Rogers describes that he now looked very similar to the Human Torch from the *Fantastic Four* Comic Book series. As he was rolling down the hill trying to put the flames out, Mashburn was helping by beating him with an entrenching tool (a small fold out shovel).

In the end it's not certain what hurt worse, the second degree burns on Rogers' face or the multiple blunt trauma wounds on his ribcage and torso. After getting the flames under control Rogers was no longer on fire; just smoldering. It was now time to make an assessment of the damage. For starters, the flames had burned the pocket flaps right off his jacket. One entire side of his uniform was gone. Rogers describes that his eyebrows looked like remnants of a used *SOS Brillo* pad; most of his hair was gone and what wasn't, was still smoking. The pain was incredible, but he didn't tell anybody because we had a mission to complete. Mashburn later explained: "Well Rogers, I figured you was in quite a fix, you being on "faar" and all so I grabbed the first *thang* handy and commenced to help." After this, he promised Mashburn that he would NEVER ask for his help again.

When Lt. Garrison finally came back from his recon the Rangers attacked the objective and headed back to the destroyer anchored off shore. Once back on board the ship the rest of the Rangers got a really good laugh at Rogers' predicament. As the Destroyer was pulling back into harbor, Garrison had them all standing up on deck in formation. As he walked in front of Rogers' squad he finally noticed that one of his men looked like an extra from *Apocalypse Now*. "Rogers, what in the world happened to you?" "Sir, I can explain. You see sir, I was just checking our grid coordinates when my poncho caught fire and Mashburn beat me with his entrenching tool so as not to cause me any further damage – although I think two of my ribs are broken." Garrison raised his hand for him to stop explaining. "Forget I asked. And Rogers, you will pay for those unauthorized alternations to your uniform. Is that clear?"

Even though his burns healed and his ribs mended, Rogers says he hasn't been able to look at a candle or a shovel the same way again! He jokes that he still get shivers when somebody lights a birthday cake. The moral of the story he offers is to – always, always have fresh batteries and NEVER use a candle – for anything!

The Rangers left the objective area on a Canadian C130 enroute to Edmonton, Alberta. Along with their basic equipment the Rangers took a pallet of blank ammunition with the directive to use it or leave it in Canada. While in Canada one of the missions was to support a pilot course the Canadian Airborne Regiment was conducting.

At that time the Canadian Airborne Regiment was composed of 1 and 2 Commando. The Regimental Headquarters was in Edmonton with 1 Commando on the first floor and 2 Commando on the second floor. The Rangers conducted two missions to support the course. One exercise was on the west coast supported by Canadian warships and the other one was in a training area south of Edmonton. On one of the exercises the Rangers made was into Alberta and the weather did not cooperate. During this exercise the Rangers knew the Canadians who we would be aggressing had Ferret Scout cars, so they wanted to be as light as possible.[6] Going light as possible was probably an error for it got cold and started to snow and they quickly became some really miserable Rangers. The Rangers were only footed with steel-shanked jungle boots. A lesson learned early on in this deployment was to ensure the appropriate environmental gear was in the Ranger inventory in the future.

It was a wet heavy snow. By night fall the Rangers were soaked and freezing. They had been moving fast all day trying to stay clear of the "bad guys." As told they crawled into the middle of this large brushy area then set out observation posts (OP) in the four cardinal directions. To beat the cold they started some warming fires and then made certain that the OPs were just outside of the sight of the fires, that way they were certain the enemy would not see them. One of the troops took his boots off that night and one of them caught fire and the darn thing shrunk until he could not get his foot back in it. He thought a little then cut the toe of his boot out and wrapped the entire end of his boot in 100 mile an-hour-tape. Ingenious Rangers making it happen and certainly made for a few laughs.

Between exercises the Rangers conducted one of the first of many equipment demonstrations they would become famous for in the military community over the years. This was a form of Abrams Charter for they demonstrated their abilities and spread that talent to those who usually would watch with awe. That day a featured event was demonstrating how fast an M60 machine gun barrel could be exchanged. In preparation for this event the Rangers conducted a lot of rehearsals firing blank ammunition. The machine guns crews were very fast on the exchange and managed to expend most of the pallet of ammo they brought to Canada.

Wagers recalls that the Rangers also had a communications layout that displayed the various radios that the Rangers operated. As the RTO, LeMauk was selected to do the communications demonstration. LeMauk had the battalion headquarters in Fort Stewart on the radio showing off the capability of the Portable Radio Communications (PRC) 74B.[7] According to Wagers, LeMauk stood tall and proud at parade rest talking about those radios like he invented them and really impressed the Canadian Regimental Sergeant Major (RSM). After the incredible demonstration by LeMauk, the RSM asked Wagers what was his rank and he replied he was a Private! He then asked Wagers when would he get promoted and Wagers replied "as soon as he learned about radios."

Of course Wagers was kidding for LeMauk was as professional as any private in the US Army at the time and soon thereafter he went to Ranger School and then to the US Military Academy at West Point with a chest full of adorations to include the coveted Jump Wings, Ranger Tab, Expert Infantryman's Badge (EIB), and Canadian Jump Wings. At this writing he is the US Army Europe and 7th Army Provost Marshall, with over 30 years' service which all began as a Ranger RTO. This is the kind of man the Rangers develop.

REFORGER 1976

Frank Magaña of B Company remembers the Ranger involvement in REFORGER their second year in existence and the Battalions interaction, or rather competition with, the 101st Airborne Division while in Europe. The irony behind this is that the 101st had been at odds with the Rangers since early 1974 when Leuer, having just left command there, tried to recruit from the Screaming Eagle ranks which caused a stir.

In the annual REFORGER exercise conducted in September 1976 the 1st Battalion (Ranger), 75th Infantry was pitted against the 101st Airborne Division (Air Assault). The 'Screaming Eagles' where out there somewhere in the Ranger Area of Operations (AO) in southern Germany. The Ranger mission was a textbook 'US Army Ranger Battalion' mission; find, fix and destroy the enemy. Ron Rokosz had left the Rangers the year earlier to attend Graduate School at Purdue University in preparation for fulfilling the first of many Abrams Charters requirements by spreading his new found Ranger experiences to the cadets of West Point where he would eventually teach in the Department of Social Sciences.

Rokosz had relinquished command of Company B, to Captain Jeff Ellis (who eventually retired as a Colonel) who deployed the company on the REFORGER mission. For this

Rangers patrolling in Germany on REFORGER. (Source: Currie)

mission, Ellis was given a company reconnaissance sector of responsibility. The company sector was further broken down into platoon sectors and subsequently down into Squad sectors.

The 1st Platoon Leader, 1LT Doug Dountz, moved the platoon to the edge of the reconnaissance sector and gave a quick Warning Order. The weapons squad leader, SSG Reggie Salinas, attached one machine gun crew to each rifle squad and Salinas himself was attached to the 1st Squad led by SGT Ron Fallon. As the squad fire team leaders prepared their fire teams for the mission the squad leaders assisted Dountz in developing a platoon reconnaissance (recon) plan. Within the hour Dountz issued a quick operations order complete with squad recon sectors, platoon rally points, rendezvous points, and a few important contingency plans.

First Platoon's squads headed out into their sectors of responsibility on their respective recon missions. Salinas and Fallon had worked together for about two years; they were both technically and tactically proficient and this would prove a perfect stage for what was to come later that evening. It was late afternoon and the 1st Squad was a few hours into the recon patrol using the fan reconnaissance technique when they took a short break.[8] Shortly after the squad removed their rucks and settled behind them in a tight security perimeter, they could hear in the distance an ever-so-slight humming or grumbling. Not recognizing the sound, the squad cut the break short and readied themselves for movement. Remaining within their sector the squad cautiously changed direction and set out to reconnoiter the area where the sound was emanating.

The humming grew louder and louder as they continued to move. It was early evening now and growing darker by the minute, the type darkness that one can only witness in

Rangers led by Sergeant Nels Nelson on a training exercise in Alaska 1977. (Source: Currie)

a thick wooded German forest. It was nearing what is called in military jargon – end of evening nautical twilight or EENT – when the squad heard what sounded like significant vehicle activity and the now distinctive sound of large generators.

SGT Fallon radioed Dountz giving an initial Situation Report (SITREP) and continued surveillance. Using AN/PVS2 Night Vision Device (a starlight scope allowing one to see in the dark using ambient light to illuminate objects in a green haze) Fallon and SGT Nels Nelson's (retired as a First Sergeant) Fire Teams low crawled to a suitable observation point (OP).

From this vantage point, the Rangers detected many generators, many vehicles, and a few soldiers. The sound was now almost deafening as more generators were fired up. The 1st Squad knew that this was a big find, but just how big they did not know! The Fire Team pulled back from the OP to the Objective Rally Point (ORP) and radioed another detailed situation reports (SITREPS) to Lt. Dountz. The Lieutenant had already linked-up with the rest of the platoon and was heading toward the 1st Squad's ORP. That SITREP was also relayed up the chain of command. Armed with this information Capt. Ellis told the 1st Platoon to hold tight and that Company reinforcements were on the way.

Ellis mustered the remainder of Bravo Company and moved toward 1st Squad, 1st Platoon's location. The 1st Squad continued reconnaissance focusing on movement routes to the generators locating the direction that generator cables were headed. The cables were on a narrow path. The 1st platoon arrived at the ORP followed shortly by Ellis, Lt. 'Hog' Brown (retiring as a Major and a Ranger Hall of Fame inductee), the Executive Officer (XO) and the remainder of Bravo Company. The entire Ranger Company was now completely formed. Salinas gave the commander a quick SITREP and the information was passed through the Ranger column. Salinas stayed near the commander as he had one of the AN/PVS2s and he could be the commander's night eyes.

It was a no moon, chilly night, and it was just before midnight. If it had been raining it would have been a perfect night for a Ranger Raid! The Company was ready to move to the target, wherever and whatever is was. Ellis told Fallon, "Lead the Way Ranger" and 1st squad moved out with two trusted fire team leaders, SGTs Steve Hawk and Nels Nelson, at the head of the Ranger file. Hawk was on point followed closely by Fallon and Nelson and the rest of the squad. They followed the generator cables up a narrow trail as they led Bravo Company toward the enemy objective.

As 1st Squad slowly moved up the generator cable trail Hawk suddenly bumped his head on something hard. As Hog Brown would say, "it was harder than a footlocker full of wood-pecker lips." Hawk reached up and felt a long thick barrel of what turned out to be a 20 mm Vulcan Gun.[9] Hawk was then 'challenged' for the 'password' by an enemy guard who was yet unseen. Thinking quickly, Hawk gave the 'challenge' back to the guard. Fallon then called for Hawk to give him the 'password.' Hawk gave the 'challenge' word again and Fallon said "Hawk, that's the 'challenge' what's the 'password.' He said, "SGT Fallon, I forgot the pass-word." Fallon then started to reprimand Hawk for forgetting the password as Fallon moved a bit closer to the guard. Fallon then took the guard to the ground with a rear strangle take down. The encounter lasted fewer than ten seconds. The enemy guard was bound with rope and thrown into the back of the Vulcan vehicle where another guard was deathly afraid to come out. The 1st Squad continued leading the Company by following the generator cables through a multi-layered defensive perimeter.

Confronted by another layer of defense, Hawk 'challenged' the soldiers and they gave the Rangers the 'password.' The enemy sentries were then promptly captured, bound, and gagged. Just then a jeep in 'black out drive' pulled up.[10] Fallon made a quick plan with Nelson for each of them to 'challenge' the enemy in the jeep and to shine their flash lights in the faces of the two in order to ruin their night vision. If any Screaming Eagles saw the Ranger uniforms, they would know that they were not 101st soldiers. The enemy soldiers in the jeep were 'challenged' and the flashlights shined in their faces. The jeep driver was especially irate about the lights in the face, but gave the 'password' and they were allowed to pass. They were more than likely Military Police (MP). Bravo Company was still in file coming up the generator path.

As Bravo Company moved forward, the squad and fire team units of the entire Company were conducting similar operations disarming and capturing Vulcan Guns, vehicles, personnel tents, supply tents, radio trucks, and generators. SGT Joe Mattison (came to the Rangers as a Private First Class and retired as a Command Sergeant Major) captured a 'Rat Rig' radio truck.[11] Thinking the stairs were located where they should have been, he went to charge up the stairs to the door of the vehicle, but was almost knocked out by the tail gate because the stairs were not in the down position. Mattison's fire team hopped up on the truck and took over the radio equipment, silencing all radio communication.

Scanning the area with his AN/PVS2, Salinas saw some lights off in the direction of what appeared to be a General Purpose (GP) Large tent. He handed the Night Vision Device (NVD) to Ellis who remarked that "the place is lit up like a Christmas tree." Fallon's Squad was now securing a few enemy positions and SGT (now promoted from Specialist Fourth Class when first introduced earlier in this work) Frank Magaña, another Squad Leader in 1st Platoon and his Fire Team Leader Don Carroll, were now near the front of the Ranger column with Ellis. Magaña's Squad moved toward the lights of the tent and hit a few strands of concertina wire. Ellis knew that a tent with generator cables leading to the tent and concertina around the tent was a very good target. It must have been some type of Headquarters.

Magaña's Squad entered the tent followed by Ellis. The Commander, .45 caliber pistol in hand said, "Gentleman, put your hands up, you have been captured by Bravo Company, 1st Ranger Battalion." The tent was the 101st Division Tactical Operation Center (TOC). Within minutes, the 1st Platoon, supplemented by the XO, Hog Brown, and parts of other platoons secured the Screaming Eagles TOC. By all accounts, it was a sight to see.

In the tent there were a few 101st officers sleeping, no doubt the G3 Operations Officer and G2 Intelligence Officer; there were gigantic map boards; radios, field phones, tables, chairs, and cots. The 101st Division Tactical Operations Center, the command and control center of the entire Division, was captured! Every 101st 'friendly' position and every suspected 'enemy' position was on the Situation Map. The Eagles were indeed screaming that night!

Ellis told Magaña and the others to erase the situation map and collect all maps, charts, operations orders, the Communications-Electronics Operating Instructions (CEOI), or the code book containing frequencies, authentications and other things necessary to communicate internally and anything that was of intelligence value. Ellis then ordered the recording of all radio frequencies and to then scramble the frequencies on all of the radios. Posted on a board was the intelligence officer's query of where was the location of the 1/75 Rangers? On it Hog Brown wrote large and bold "HERE." Outside the Operations tent, Rangers were disconnecting generator cables and cutting large sections from others. The Maneuver

Controllers (those that maintained control of the exercise) were awakened and took control before any fist-to-cuffs could break out. The TOC was officially captured.

Unfortunately, the Rangers did not find the Commanding General.[12] They later found out his tent was a few tents down from the TOC. Regardless, the raid was successful and the victorious Rangers were allowed to leave the enemy perimeter unhindered. It was now nearing BMNT (begin morning nautical twilight) and B Company, 1st Ranger Battalion was on the move to their next mission. In the end, the maneuver exercise was stopped for two days while the 101st Division attempted to repair their severely damaged TOC, contact their units and reestablish communications. The Rangers proved themselves to be a force to reckon with that night.

In the years to come, Rangers were involved with every operation leading up to the War on Terror which will be covered in Chapter 6. The remainder of this chapter will cover all combat operations involving Rangers from the Iranian Hostage rescue attempt in 1980 to the aborted Haiti Mission in 1994. The likes of KC Leuer and the original plankholders have now been replaced by other great Americans. Leuer went on to command a brigade in the 24th Infantry Division (Mechanized) and eventually made the rank of major general and commander of Fort Benning. There is no doubt his legacy as the original commander of the modern day Ranger has lived and will live forever in the annals of American Military History. But time marches on and the nation began to call on the Rangers that Leuer built for a series of combat operations lasting over four decades from the time of their inception. The following is a list of missions that the Rangers have been a part of since their inception in 1974:[13]

- Operation Eagle Claw which was the 1980 rescue attempt of American hostages in Tehran, Iran.
- In 1983, the 1st and 2nd Ranger Battalions conducted Operation Urgent Fury, a combat parachute assault onto the island of Grenada.
- All three Ranger battalions, with a headquarters element, participated in the US invasion of Panama for Operation Just Cause in 1989.
- The first platoon and Anti-Tank section from Alpha Company and B Company, 1st Battalion was deployed in the First Persian Gulf War for Operations Desert Storm and Desert Shield in 1991.
- Bravo Company, 3rd Ranger Battalion was the base unit of *Task Force Ranger* in Operation Gothic Serpent, in Somalia in 1993; concurrent with Operation Restore Hope.
- In 1994, soldiers from the 1st, 2nd and 3rd Ranger Battalions deployed to Haiti before the operation's cancellation. The force was recalled five miles from the Haitian coast.

EAGLE CLAW – BAPTISM OF FIRE

As a result of the demonstrated effectiveness of the Rangers during multiple Army, Joint, and special operations exercises in the years following their inception in 1974 the nation could now depend on the Rangers to deploy them to fight the nation's wars. Five years after the activation, the 1st and the 2nd Battalions were called to execute a risky raid in the Middle East.

On November 4, 1979 the US Embassy in Tehran, Iran was overrun. Followers of the Ayatollah Khomeini took 66 American hostages that day and over a series of days released

13. This left 53 hostages to bargain with for the release the former Shah of Iran, Mohammed Pahlevi, so he could stand trial for what they termed crimes against the state. The US, in its policy to not deal with terrorists and their demands, refused and in response imposed economic sanctions.[14]

In Iran, the hostage taking was widely seen as a blow against the US, and its influence in Iran. In the United States it was perceived as an outrage violating international law which grants diplomats immunity from arrest and protection status for diplomatic compounds. The crisis has also been described as the "pivotal episode" in the history of Iran–United States relations."[15]

The crisis also marked the beginning of US legal action with the economic sanctions against Iran which further weakened economic ties between the two countries. In the decades to follow the crisis has been described as an "entanglement of vengeance and mutual incomprehension"".[16] Finally, it was determined that it was a matter of national honor as well as moral and political obligation to rescue the remaining 53 captives. After a series of failed negotiations for their release, President Jimmy Carter authorized a military rescue mission on April 11, 1980 with a target date of the 24th to execute the President's orders.

The rescue operation was codenamed Operation Eagle Claw. The plan called for an absolute minimum of four helicopters to successfully conduct the operation. Once the plan was developed leaders decided that six helicopters ensured of its success. Military leaders then decided to use eight to reduce the risk of failure.

In its eventual execution two of the helicopters could not navigate through a very fine sand cloud which forced one helicopter to crash land and the other to return to the aircraft carrier USS *Nimitz* (CVN-68). Six helicopters reached the initial rendezvous point, which was named Desert One, and often is used in lieu of Eagle Claw as a naming convention for the mission. One of the helos had damaged its hydraulic systems and the spares were on one of the helicopters that aborted. In the early planning stages it was determined that if fewer than six operational helicopters were available, then the mission would be automatically aborted, even though it was decided that only four were absolutely necessary for the operation.

In a controversial move, the commanders on the scene requested to abort the mission. President Carter gave his approval. As the force prepared to leave Iran, one of the helicopters crashed into a C-130 Hercules transport aircraft containing fuel and a group of soldiers. The resulting fire destroyed the two aircraft involved and killed eight Americans.[17]

A RENDEZVOUS WITH DESTINY

Eagle Claw was one of the first missions conducted by Delta Force partnering with US Army Rangers. The complex operation was designed as a two-night mission with the first stage of the mission delivering a mostly Ranger qualified Delta Force rescue team to a small staging site, near Tabas, Iran in Yazd Province.[18] The Desert One site was to be used first as a temporary airstrip for the three US Air Force special operations MC-130E Combat Talon I penetration/transport aircraft and three fuel bladder equipped EC-130E Hercules.[19] It was then to become a staging base for eight Navy RH-53D Sea Stallion minesweeper helicopters flown by US Marine Corps personnel off the aircraft carrier USS *Nimitz* (CVN-68) in the nearby Indian Ocean. These aircraft were to transport the rescue team to Tehran.

The sequence was to be initiated with the C-130s flying in under the radar and landing at Desert One to off-load men and equipment and refuel the arriving helicopters, which would, in turn, execute the rescue operation. The helicopters would fly the ground troops to a hide site called Desert Two near Tehran where the helicopters would be concealed. The next night, the rescue force would be transported in trucks to the embassy by CIA agents, overpower the guards, and escort the hostages across the main road in front of the embassy to Shahid Shroud Stadium, where the helicopters would retrieve the entire contingent.[20]

With an AC-130 gunships orbiting overhead the helicopters were to transport the rescuers and hostages from Shahid Shroud Stadium to Manzariyeh Air Base outside of Tehran. The goal was for a Ranger force to seize the airfield at Manzariyeh to permit C-141 transports to land in advance of the rescue to transport the contingent from Iran under the protection of fighter planes.[21]

The participants in this operation where successful special operations leaders. The Joint Task Force commander was Ranger qualified US Army Maj. Gen. James B. Vaught, the fixed-wing commander was Col. James H. Kyle, the helicopter commander was Marine Lt, Col. Edward R. Seiffert, and Delta Force commander was the legendary Col. Charlie Beckwith.[22]

EXECUTING EAGLE CLAW

In the eventual execution of the mission the delivery of the rescue force; the special operations aircraft MC-130E Combat Talons with the call signs of *Dragon 1* to *3* and the EC-130Es with call signs *Republic 4*, *5* and *6*, went according to plan. The special operations transports took off from their staging base at Misirah Island near Oman (the same launch point that would be used for the Ranger invasion of Afghanistan in 2001) and were refueled in-flight by KC-135 tankers just off the coast of Iran. The mission called for a group of 130 Delta, Rangers, drivers, and translators to be inserted into the Iranian desert.[23]

Dragon 1 landed at 22:45 local time under blacked-out, visible only through night vision goggles. With significant damage to the aircrafts wing upon landing it off-loaded a USAF combat control team (CCT) led by Carney, Beckwith and part of his 120 Delta operatives, 12 Rangers of a roadblock team, and 15 Iranian and American Persian-speakers, most of whom would act as truck drivers. The CCT immediately established a parallel landing zone north of the dirt road and set out beacons to guide the helicopters. More aircraft landed and discharged the remainder of the Delta Force operatives and the fix wing departed to make room for eight RH-53D Sea Stallion helicopters; with call signs *Bluebeard 1* to *8*.

Bluebeard 6 was grounded and abandoned in the desert when the pilot determined it had a cracked rotor blade. Its crew was picked up by *Bluebeard 8*. The remaining helicopters ran into an unexpected weather phenomenon known as a *haboob*.[24] *Bluebeard 5* flew into the haboob but abandoned the mission due to visibility challenges and returned to the *Nimitz*. Aircraft issues continued to plague this mission. *Bluebeard 2* arrived at Desert One with a malfunctioning second-stage hydraulics system leaving one hydraulics system to control the aircraft.[25]

As the first on site began securing Desert One, a tanker truck apparently smuggling fuel was blown up nearby by a shoulder-fired rocket as it tried to escape the site. The passenger in the tanker truck was killed, but the driver managed to escape in an accompanying pickup truck. The resulting fire illuminated the area for many miles around, and provided a visual guide to Desert One for the incoming helicopter crews.[26]

With only five Sea Stallions remaining to transport the men and equipment to Desert Two, which Beckwith considered was the abort threshold for the mission, the various commanders reached a stalemate. The air commander, Seiffert, and the ground commander, Beckwith, learned an important lesson on this mission which has been infused into special operations and conventional air mission command training and planning for years to come. This lesson was to war game the "bump plan" so that the aircraft and troop requirements are always synchronized and abort criteria are built in to the mission development process. In the end, after only two and a half hours on the ground, the abort order was received.

RANGERS IN ACTION

As stated, the overall mission commander was Col. Charlie A. Beckwith. A veteran of Korea and Vietnam, "Chargin' Charlie" was a hardened Green Beret Special Forces and a Ranger qualified officer who was considered to be the premier US expert in unconventional warfare. Beckwith was commander of 1st Special Forces Operational Detachment-Delta (1SFOD-D) more commonly referred to as "Delta Force," at the time a secret elite team of commandos who were specifically trained in a number of antiterrorist tasks, one of which was to surreptitiously infiltrate target areas dressed in civilian clothes and free hostages from buildings.

If all went right, and we now know it didn't, eighty Rangers would be airlifted from the Egyptian Airbase at Qena to an isolated desert airstrip at Manzariyeh. Located thirty-five miles south of Tehran, this airstrip was part of an unoccupied former bombing range and had an asphalt paved runway that would be secured by the Rangers and used by C-141 Starlifters.

Withdrawing from the embassy compound to the airstrip by helicopter, the Delta commandos and hostages would load the Starlifters. Then they, along with the helicopter crews, drivers, translators, and DOD agents as well as select individuals who were operating in Tehran in support of the rescue attempt, would depart for Qena, mission complete.

Once the other elements were lifted out, the Rangers would collapse their perimeter and then depart. Delayed demolition charges would ensure the destruction of the helicopters left behind.

While not executed as planned, the mission was indeed launched on April 24 and the majority of the ground forces, to include a detail of Rangers from C Company, 1st Battalion, 75th Rangers commanded by then-Capt. David L. Grange would later command the Ranger Regiment and retire as a General officer. Under Grange's command they boarded two C-141s and flew to Misirah Island. Later that evening, they departed for Desert One.[27]

Keith Nightingale was Gen. Vaught's deputy operations officer. Nightingale was charged as the Ranger and US Army proponent with getting the Rangers on the mission. Whether it was parochial or just personal, the Rangers inclusion into the mission in Iran was fought vigorously by Beckwith who wanted no one other than Delta operatives involved. However, it was very clear to Vaught that Rangers had to be engaged. In reality Delta didn't have the force available required to do anything but the direct action missions.

Someone had to do the non-direct mission sets such as those required to provide security at Desert One, security at Desert Two, security at the extraction airfield and security at a potential takedown of a place called Nain. Finally after a discussion in Chief of Staff of the Army Edward C. "Shy" Meyer's office in which Nightingale was present, Beckwith agreed to the Rangers inclusion, for the alternative was reportedly that he be relieved. Beckwith

A generation of Rangers: The courageous and well renowned Grange Ranger Family
(FORT BENNING, Ga.) Lt. Gen. David E. Grange Jr., whom the Best Ranger
Competition is named after, and his son Maj. Gen. David L. Grange attend the Ranger
School Graduation of 2nd Lt. Matthew C. Grange, Friday, December 7, 2012 at Victory
Pond. Also in the photograph: 2nd Lt. Grange's maternal grandfather, Brig. Gen. (retired)
Charlie Getz. Getz commanded the Florida Ranger Camp and is the recipient of the
nation's second highest combat award, the Distinguished Service Cross (DSC).
(Photo by: Ashley Cross/MCoE PAO Photographer) (Source: Grange)

Captain Dave Grange with fellow Rangers at
Hunter Army Airfield (HAAF) in Savannah,
Georgia in 1978 to include then-Capt Dell Dailey
who went on to retire as a Lt. Gen. and Director
of the Center for Special Operations and then as
an Ambassador at large and the US Department of
State's coordinator for counterterrorism.
(Source: Grange)

insisted that if the Rangers were included then they had to report to Jesse Johnson-a member of Delta support staff-which ended up happening.

Grange was the commander of the C Company, 1st Rangers which formed the nucleus for the Ranger package. This mission set the standard for the now common Delta-Ranger donut defense where the Rangers secure the outer ring and Delta performs the missions on the inner. In preparation, Rangers executed airfield takedown and security training as well as a lot of demolitions work where the Rangers were given the contingency to blow the helicopters at the extraction airfield as required.

During the mission the Delta support team prepared to receive arriving helicopters on the airfield. To support this, the Rangers established security on the routes in and out of the airhead. They had no sooner off-loaded a jeep to support their task when they made contact with a Mercedes bus that entered the airfield. Halting the bus with warning shots, the forty-five distressed Iranians were unloaded and secured. Minutes later, another approaching vehicle was spotted; a small fuel truck. Failing to halt, the vehicle was set afire by an anti-tank rocket fired by a Ranger. Quickly exiting the burning truck, the driver ran a distance to a pickup that had moved into the perimeter undetected and the pickup escaped in a hail of bullets. The Rangers maintained security of the airfield as the aborted mission aircraft attempted to depart the area.

The ultimate task was to close up Desert One and get everyone safely back to Egypt. In the early mornings the next day a helicopter on the airfield was maneuvering to top off with fuel from a C-130. The pilot became disoriented in the great swirls of dust created by his engine and that of the C-130. Moving left, then right, the helicopter banked and crashed into the refueling C-130, creating an explosion and a tremendous fireball in the desert night. Thirty-nine soldiers on the C-130 were evacuated from the damaged C-130. The five Air Force crewmen in the C-130 cockpit as well as three Marines in the helicopter perished in the flames and four Army soldiers suffered serious burns.[28]

Believing that the force was in imminent danger Kyle ordered the abandonment of all helicopters for the entire force of Delta, Rangers, and support personnel to load on the remaining C-130s. Equipment was jettisoned from the C-130s to make room for the extra bodies that had not been originally manifested for the aircraft.

Shortly after the first failed, planning and training for a second rescue mission attempt was authorized under the name *Project Honey Badger* and accompany operational name of *Operation Credible Sport*. However, the impending change of administration in the White House forced the abandonment of this project. In the end, the mission to rescue the American hostages was left behind on Desert One. Left behind at Desert One were five serviceable helicopters, weapons, communications equipment, secret documents and maps, as well as fifty three hostages. A second mission was never attempted. Having served as hostages for 444 days, the 53 Americans were released on January 20, 1981, just minutes before Ronald Reagan was sworn in as the President of the United States.[29]

While the first attempt to integrate US Army Rangers with Tier 1 special operations forces was not a success it set the stage for the special operations community to perform to standard on future missions. As a result, Delta Force, the Rangers and other Special Operations units would prove so successful in the future that the nation relied upon this force more and more as the missions became more complex and the requirement for success nonnegotiable.[30]

In 1984, the Department of the Army announced the activation of a third Ranger Battalion. As a result, the 3rd Battalion, 75th Infantry (Ranger), and Headquarters and

Plankholder Keith Nightingale as a Ranger Commander in Grenada 1983.
(Source: Nightingale)

Headquarters Company, 75th Infantry (Ranger), received their colors on October 3, 1984, at Fort Benning, Ga. On February 3, 1986, World War II Battalions and Korean War Lineage and Honors were consolidated and assigned by tradition to the 75th Ranger Regiment making its formation and activation complete. Not since the Second World War has the Rangers assembled such a large force in the ranks of the US Army. Additionally, this marked the first time that an organization of that size organized and recognized as the parent head-quarters of the Ranger Battalions.[31]

URGENT FURY, GRENADA

Gaining its independence from the United Kingdom in 1974, Grenada's constitution was suspended by a leftist New Jewel Movement some five years later in a 1979 military coup. With the 1983 murder of Prime Minister Maurice Bishop, the US responded quickly to the aid of the island nation.

Located 100 miles from the shore of Venezuela, South America, Grenada was the victim of an October 13, 1983 bloody military coup led by former Deputy Prime Minister Bernard Coard which had ousted the four-year revolutionary government. The plan and eventual outcome for the US invasion was to restore its constitutionally elected leaders.

Violence and hardline Marxism caused deep concern among neighboring Caribbean nations, as well as in Washington, DC. Also, the presence of nearly 1,000 American medical students on the island caused added concern for the Reagan administration.[32] Additionally, the unrest provided a timely reason for the US to eliminate the threat from the Soviet Union and other Communist efforts in the region, such as Cuba.

Exiting the Aircraft on a Combat Jump.

This was a volatile time in US international policy. The suicide bombing on the Marine Corps barracks in Beirut which claimed 240 lives just 10 days after the coup in Grenada added importance to a successful reaction by the US in this instance. Claiming American imperialism, international media made any reaction to the coup controversial. Regardless, the Organization of American States (OAS) and the Organization of Eastern Caribbean States (OECS) supported intervention to expel Soviet and Cuban presence on the island and the holding of American medical students at the True Blue Medical Facility and condemned the internal rife supported by the Soviets as "a flagrant violation of international law."[33]

COMBAT JUMP

Abrams' farsightedness and of trust in the combat effectiveness of the Ranger battalions was validated during Urgent Fury. The mission of the Rangers was to protect the lives of American citizens and restore democracy to the island. To do this the Rangers first had to get on the tiny island. On October 25th, some twelve days after the coup, the 1st and 2nd Ranger Battalions conducted a daring low-level parachute assault at approximately 500 feet to seize the airfield at Point Salines and then move to and rescue American citizens at the True Blue Medical Campus.[34]

A rapid deployment force consisting of the 1st and 2nd Ranger Battalions, the 82nd Airborne Division, a contingent of US Marines, Delta Force, the 160th Special Operations

Aviation (Night Stalkers) and Navy SEALS made up the 7,600 troops from the United States, Jamaica, and members of the Regional Security System (RSS) were involved in this mission. The Rangers conducted the airborne drop on Point Salines Airport on the southern end of the island while a Marine helicopter and amphibious landing occurred on the northern end at Pearl's Airfield shortly afterward.[35]

The 1st Battalion out of Hunter Army Airfield (HAAF), Georgia commanded by Lt. Col. Wesley Taylor was assigned the tasks of securing the Point Salines Airfield and the True Blue medical school campus occupied by US citizens at the eastern end of the runway. The 2nd Battalion, commanded by Lt. Col. Ralph Hagler from Fort Lewis, Washington was assigned the mission of seizing Pearls Airport. After the first volley of parachutists landed, President Reagan said "we got there just in time" indicating that American medical students were in grave danger before the Rangers could quickly rescues them."[36]

One of the major challenges recognized early on with the invasion is that little was known about the enemy the invading force faced, for there was no tactical intelligence or even details of locations. Military maps of the island did not exist and much of the planning was based on an old, outdated British 1:50,000-tourist map of the region. As a result, black and white photocopies of this map were distributed for planning purposes.

Defending the island was the People's Revolutionary Army (PRA) which was believed to be 1,200 strong with an additional 2,000 to 5,000 militia and 300 to 400 armed police to reinforce as needed. Soviet antiaircraft 12.7-mm machineguns and eighteen ZSU-23 37 mm anti-aircraft cannons were in the inventory, as were eight BTR-60 Armored Personnel Carriers (APC). There were also a small number of 75-mm antitank recoilless rifles and 82-mm mortars.[37]

The plan was for 600 Rangers from both Ranger Battalions to land or drop, depending on the conditions encountered at the Salines airfield. 1st Battalion, in addition to securing the airfield and True Blue, would provide a company of reinforcement to the sixty Special Operations soldiers for the St. George area of operations. The mission of the 2nd Battalion was to attack the PRA base at Calivigny.

H-Hour for the jump was set for 0200 on October 25, however this would be changed to 0500. At 2130 on October 24, the first MC130s carrying Rangers took off. A half-hour after departure, the Rangers were informed that the runway was definitely blocked. At 0330 the Pathfinder team jumped at 2000 feet above the ground (AGL) from the reconnaissance C-130 flying on station over the port. Once on the ground, the Pathfinders confirmed that heavy equipment was blocking the runway. As a result, the Rangers were going to have to jump to accomplish the mission. By 0400, the order was given that all Rangers were to chute-up for a jump and quickly performed in-flight rigging to meet the 0500 hit time.[38]

In chalk five, the executive officer for 1st Battalion, Maj. Jack Nix (a retired Lieutenant General) anticipated the jump and directed his men to rig their chutes. However, a series of chaotic exchanges between the jumpmasters and Air Force loadmasters caused confusion as to the decision to jump or land. This chaos was replicated on transports six and seven and issues continued to multiply for the airborne assault element. With only fifteen minutes left before the jump navigational difficulties resulted in the runway clearing team aircraft to race track in behind the assault elements chalks dropping them out of order.[39]

Taylor was infuriated for he was now dropping his assault element before the runway could be cleared for follow-on airlands. As a consequence, H-Hour was pushed back thirty minutes to 0530. Almost daylight, the sky was partly cloudy and the winds were windy at twenty knots. From an altitude below the minimum safety altitude of 500 feet AGL, the first Rangers exited on the green light at 0534. A 12.7-mm four-barrel antiaircraft machinegun opened up, spewing green tracers across the sky. It wasn't until 0710, some 90 minutes later that last Ranger hit the ground.[40]

On the ground, the 1st Battalion secured the True Blue Medical School campus, the airfield's control tower, the area surrounding the terminal and the village of Calliste. The 2nd Battalion was able to assemble quickly and met little resistance. They were able to deliberately clear the runway area west of Hardy Bay, clear the narrow strip of land south of the runway towards Point Salines and secure the low hills from the north down to Canoe Bay.

During their movement into the village the Rangers suffered their only killed in action (KIA) on the runway when an M-60 gunner, Private Mark Okamura Yamane, was struck by a round in the neck after calling out in Spanish to an enemy force to surrender. Sergeant Manouos Boles then proceeded to lead the attack against the entrenched enemy by driving a captured dozer up a hill toward the school. A feat that became famous in the annals of Ranger history.

The brief firefight on the runway came from a Cuban construction camp. Fidel Castro, the Cuban dictator, provided arms and approximately forty Cuban advisors and 635 armed, but mostly unfit, Cuban construction workers in support of Grenada's totalitarian regime who mostly broke ranks and ran during the invasion.

Securing the high ground northeast of the terminal, the Rangers were able to look down into the Cuban headquarters at Little Havana and observe two mortars in action. The Rangers fired on the Cubans using a captured 12.7 mm machine gun and the Cubans quickly scattered leaving behind their weapons. By 1000 that day Calliste was secure.[41]

TRUE BLUE

The Ranger platoon assigned to rescue the medical students arrived at the True Blue Medical Facility and secured it in fifteen minutes later. With the arrival of the Rangers the PRA soldiers fled north into the hills. The students there unharmed and the Rangers suffering no casualties in the assault. However, the Rangers quickly recognized that only half of the students were present with remainder at a campus called Grand Anse.

While the raid at the True Blue campus had been a success, the same could not be said for when C Company, 1st Battalion and members of Delta Force attacked the Richmond Hill Prison. Unable to locate the prison at first, and eventually finding it atop a high ridge covered in dense foliage, the attacking force quickly realized that the only way into the prison was for the Black Hawks to hover over it allowing the Rangers to fast-rope in.

Unknown to the Rangers there was a military base known as Fort Frederick sitting on a ridge only 300 meters across the valley to the east of the prison. The fort atop the ridge dominated the prison below it by at least 150 feet. And, in actuality, it was not the fort but the two antiaircraft guns that would make the air assault risky. The prison was a tough enough objective to get into without the overwatching and unanticipated heavy weapons.

Flying low over the valley and attempting to evade enemy fire the pilots attempted failed to locate a place to put the force down in the prison. On the second run, the pilot of the fourth Black Hawk, Capt. Keith J. Lucas, who had already suffered a wound to the right arm on the first go-around, was killed when five rounds fired from above smashed through the windshield and struck him in the head and chest. Though suffering a grazing wound himself, the copilot, CWO2 Paul Price, was able to maintain control of the seriously damaged and burning aircraft.

Escorted by a second Black Hawk, Price flew south towards Point Salines and struggled to keep the battered helicopter airborne. Unable to maneuver, Price was forced to fly over the PRA base at Frequente. Hit again, the Black Hawk crashed around 0640 just on top of Amber Belair Hill with such force that the aircraft burst into flames. Despite the intensity of the flames, the severity of the impact, and the number of machinegun rounds that had passed through it, all of the crew but Lucas and the Rangers on board survived.

In an attempt to locate the remaining 230 American medical students and faculty still unsecured, the 2nd Battalion departed on nineteen helicopters to make the short six-kilometer flight to Grand Anse. Here they easily secured and extracted the American non-combatants giving closure to the strategic objective of securing American lives.

The final mission remaining for the Rangers was Calivigny where intelligence estimates placed the garrison strength at 600. On the 27th pre-assault fires hammered the camp before four flights of Black Hawks, each loaded with approximately fifteen Rangers made their way to the objective but were raked with enemy fire. Within twenty seconds three UH-60s were destroyed and three exposed Rangers from one of the flights were killed and another four seriously wounded. The travesty of it all was that the mission had been all for naught because once they landed the Rangers discovered that the garrison was unoccupied. The two Ranger battalions redeployed to their home stations on October 28. The fight for Grenada was over.

In the end, the military government of Hudson Austin was deposed and replaced by a government appointed by Governor-General Paul Scoon. This national leadership remained in place until elections could be held in 1984.

The date of the invasion is now a national holiday in Grenada called Thanksgiving Day. The Point Salines International Airport was renamed in honor of Prime Minister Maurice Bishop.

The invasion highlighted issues with communication and coordination between the branches of the United States military, contributing to investigations and sweeping changes, in the form of the Goldwater–Nichols Act and other reorganizations.[42] It also highlighted the capacity of the Rangers and the ability to deploy quickly and accomplish the job in a joint environment. What stood out to the senior leaders was the flexibility of the Ranger leaders and their cohesive units to react quickly and adjust literally while in flight.

Operation Urgent Fury was the first battalion level combat mission planned and executed by the Rangers since the end of World War II, some thirty-eight years earlier. It wasn't perfect by any means, but the modern day Ranger would now be used in every combat deployment to come.

THE RANGERS EXPAND THEIR RANKS

Following the Grenada invasion the senior military leaders decided to add a third Ranger battalion to the ranks activating the 3rd Ranger Battalion, 75th Ranger Regiment on October 3, 1984 at Fort Benning, Georgia. The former Army G1, Lt. Gen. John Le Moyne, who was a member of 2nd Ranger Battalion, stated that the activation of the 3rd Ranger Battalion was actually delayed due to a lack of soldiers of high enough caliber. Le Moyne said that the Army never waived from the decision not to accept a lesser caliber of soldier just to meet a programmed force structure.[43]

On May 10, 1984 General John Wickham, the Chief of Staff of the US Army (CSA) provided Lt. Col. Wayne A. Downing to command this battalion. Downing would go on to command the Regiment when it was formed in 1986 as the first Colonel of the Ranger Regiment. To control the three Ranger Battalions a higher headquarters was needed. On February 3, 1986 the Secretary of the Army, John O. Marsh, Jr. announced the activation of the 75th Ranger Regimental Headquarters at Fort Benning, Georgia. To meet the hierarchy of command and control of the three battalions a headquarters was needed. Upon activation the CSA provided the following guidance:

> The Ranger Regiment will draw its members from the entire Army – after the service in the Regiment – return these men to the line units of the Army with the Ranger philosophy and standards. Rangers will lead the way in developing tactics, training techniques, and doctrine for the Army's light infantry formation. The Ranger Regiment will be deeply involved in the development of Ranger doctrine. The Regiment will experiment with new equipment to include off the shelf items and share results with the light infantry community.[44]

As the 35th Anniversary of the Korean War-era Rangers was ongoing the activation of the new Ranger organization was a fitting way to tie the future of the force with the heroes of the past. Tying the lineage of the World War II and Korean War Ranger units to the activation of the new Ranger headquarters marked the first time that an organization of that size had been officially recognized as the parent headquarters of the Ranger Battalions.[45] Not since the Darby's Ranger Force era had the US Army had such a large Ranger force, with over 2,000 soldiers being assigned to Ranger units.[46]

As the Ranger organization expanded so did the training base. Abrams Charter was fast filling the ranks of not only the newly formed Regiment, but also the Big Army. On November 1, 1987 the Ranger Department; originally organized in October 1951, was reorganized after 36 years into the Ranger Training Brigade (RTB) organized into three Ranger Training Battalions.

Three Ranger training battalions, the 4th, 5th and 6th Ranger Battalions, were re-activated as the Ranger Training Brigade (RTB). The 4th Ranger Training Battalion is located at Fort Benning, Georgia. Camp Frank D. Merrill at Dahlonega, Georgia is home of the Ranger School Mountain phase which was supported by the 5th Battalion. The 6th Ranger Training Battalion is located at Camp James E. Rudder at Auxiliary Field Six in Eglin Air Force Base, Florida and home to the Florida or swamp Phase for Ranger School. These units and Ranger instructors are part of the Training and Doctrine Command (TRADOC) and are not formally included to the active strength of the 75th Ranger Regiment.

In the 1990s another Chief of Staff of the Army (CSA), General Gordon R. Sullivan, developed his own charter for the Regiment with the following:

> The 75th Ranger Regiment sets the standard for light infantry throughout the world. The hallmark of the Regiment is, and shall remain, the discipline and espirit of its soldiers. It should be readily apparent to any observer, friend or foe, that this is an awesome force composed of skilled, dedicated soldiers who can do things with their hands and weapons better than anyone else. The Rangers serve as the connectivity between the Army's conventional and special operational forces.

The Regiment provides the National Command Authority with a potent and responsive strike force continuously ready for worldwide deployment. The Regiment must remain capable of fighting anytime, anywhere, against any enemy, and winning.

As the standard bearer for the Army, the Regiment will recruit from every sector of the active force. When a Ranger is reassigned at the completion of his tour, he will imbue his new unit with Regiment's dauntless spirit and high standards.

The Army expects the Regiment to lead the way within the infantry community in modernizing Ranger doctrine, tactics, techniques, and equipment to meet the challenges of the future. The Army is unswervingly committed to the support of the Regiment and its unique missions.[47]

PANAMA INVASION

December 1989 saw the entire 75th Ranger Regiment conduct a night parachute assault onto the Isthmus of Panama for Operation Just Cause. It was the first time since its activation some three years earlier that the Regiment had been deployed to combat.

This operation also saw the integration of a new type of military leadership and oversight at the national level. The Goldwater Nichols Defense Reorganization Act of 1986 provided more authority to the Chairman of the Joint Chiefs of Staff (Gen. Colin Powell at the time) and to the various Combatant Commanders in the various regions of the world. In concept the Act was a forcing function to streamline the chain of command by reducing the hierarchy that habitually infests military commands at the mid-level.

Powell was a major stakeholder in this operation as he implemented as what would become widely known as the *Powell Doctrine*. In short, this so-called doctrine was basically a journalist-created term named after Powell when he was the Chairman of the Joint Chiefs of Staff during Desert Shield and Desert Storm in 1990-1991. In reality the principles governing this doctrine were based in large part on the Weinberger Doctrine, devised by Caspar Weinberger, former Secretary of Defense and Powell's former boss.

Regardless, what was known as The Powell Doctrine stated that the following list of queries had to be answered affirmatively before military action is taken by the United States:

Is a vital national security interest threatened?

Do we have a clear attainable objective?

Have the risks and costs been fully and frankly analyzed?

Have all other non-violent policy means been fully exhausted?

Is there a plausible exit strategy to avoid endless entanglement?

Have the consequences of our action been fully considered?

Is the action supported by the American people?

Do we have genuine broad international support?

All of these queries held true for the Panama Invasion, thus making it a model in its execution and in the war termination and conflict resolution portions of the operation.[48]

As trouble brewed in Panama, Gen. Frederick F. Woerner, Jr., the Commander in Chief of US Southern Command (USCINCSO), began developing plans to protect US lives and property; to maintain freedom of maneuver on the canal; to conduct noncombatant evacuation operations (NEO); and to develop a plan to assist any government that might replace the Noriega regime. Woerner envisioned a massive buildup of forces within US bases in Panama. The purpose of the buildup was to intimidate the Panamanian Defense Forces (PDF) leaders under Gen. Manuel Antonio Noriega with the hopes they would be motivated to overthrow him. At the very least, the buildup would provide a force capable of overthrowing the PDF.[49]

President Ronald Reagan favored the mass approach over the strategy of surprise. An en masse strategy allowed conventional and special operations forces from the United States and those already in country to conduct an overwhelming mass assault against Noriega and the PDF. One of the main goals in this strategy was to ensure the capture of Noriega to prevent him from fleeing to the hills and potentially organizing an extended guerrilla warfare. The senior US leaders also wanted to prevent the abduction or possible killing of any of the thirty-five thousand US citizens already residing in Panama.

Noriega ignored the US preparation and signals. Instead, the dictator became increasingly brutal during the time leading up to the actual invasion. For the Americans, emphasis shifted toward a plan that embodied elements of both strategies. Reagan was followed in office by George H.W. Bush who listened to Powell's advice agreeing on how to execute the mission; with elements of mass and surprise as the characteristic fundamentals.[50] Clearly a Ranger type mission.

A faction of the PDF staged a coup attempt on the *La Comandancia* on March 16, 1988. Noriega curbed the effort and purged the PDF of those he considered disloyal. He also declared a state of national emergency; cracked down on political opposition; and stepped up the harassment of US citizens, chiefly through severe travel restrictions, searches, and roadblocks.[51]

Exiting the Aircraft.

On May 7, 1989 Panamanians elected an anti-Noriega candidate, Guillermo Endara, for president by defeating Carlos Duque by a three to one margin. During the election, Noriega thugs from the Dignity Battalions tried to coerce voters into electing Noriega's candidates despite the presence of observers from the Catholic Church and former President Jimmy Carter. Noriega attributed the election results to foreign interference and annulled them. As a result Endara and other opposition leaders were physically assaulted and went into hiding finding asylum in the papal *Nunciatura*.

At this point President Bush grew increasingly concerned about the physical safety of the thousands of US citizens in Panama. In an operation called Operation Nimrod Dancer a thousand troops of the 7th Infantry Division (Light) deployed to Panama from Travis Air Force Base, California to protect US civilians. To support the operation, Marines from the 2nd Marine Expeditionary Force from Camp Lejeune, North Carolina and 762 troops of the 5th Infantry Division (Mechanized) from Fort Polk, Louisiana moved by sea and air to Panama on May 19.[52]

Opposition to the Noriega regime continued through the fall of 1989 when a US backed coup failed. The result was increased tension between US and PDF forces culminating in inflammatory Panamanian declarations of a state of war with the US because of "US aggression." This led to a series of events culminating in the killing of Marine 1st Lt. Robert Paz fueling the US Invasion.[53]

President George Bush authorized the invasion at midnight on December 20, 1989. The task was to capture Gen. Manuel Antonio Noriega and bring him back to the US to face drug-smuggling charges. As part of the mission, OPLAN 90-2, initially code-named "Blue Spoon," directed "Task Force Red," the name the special operations world gave to the 75th Ranger Regiment, to conduct an airborne assault on the Omar Torrijos International Airport and Tocumen Military Airfield. Following the airborne assault the Rangers were to neutralize the Panamanian Defense Forces' 2nd Rifle Company, the entire Panamanian Air Force and secure the airfields for the arrival of the 82nd Airborne Division. The task organization for this mission included the 1st Ranger Battalion and C Company, 3rd Ranger Battalion – designated as Task Force Red-Tango.[54]

The remainder of the regiment including the 2nd and 3rd (-) Ranger Battalions, and a Regimental command and control team, designated as Task Force Red-Romeo, conducted a parachute assault onto the airfield at Rio Hato.[55] Their task was to neutralize the PDF's 6th and 7th Rifle Companies and seize Noriega's beach house. The ready brigade of the 82nd Airborne Division, consisting of over 2,000 paratroopers, jumped forty-five minutes after the Rangers' seized Torrijos/Tocumen Airport. From the airfield the 82nd conducted battalion-sized air assaults against Fort Cimarron – home base of the PDF 1st Infantry Company, and Panama Viejo – home base of the PDF 1st Cavalry Squadron.[56]

In preparation for this operation the entire Regiment participated in a rare battalion level-regimental-size exercise conducted on December 14, 1989, just a week before the actual no-notice deployment. Named "Sand Flea", the rehearsal replicated a large airfield takedown practiced at Eglin Air Force base in Florida.[57] The exercise went to great lengths to replicate the eventual objectives with buildings being built and placed in appropriate locations to represent the actual airfields seized during the attack. Spectre aircraft from the Air Force special operations and "Little Birds" from the Night Stalkers flying in support of the mission gave the Rangers and other special operations units involved in the training a realistic feel. Most importantly, the timing and communications required for success could be synchronized during this training event mitigating the risk during the eventual operation.[58]

A NO-NOTICE JUST CAUSE

Lt. Gen. Carl Stiner, the commander of the XVIII Airborne Corps out of Fort Bragg, North Carolina, was selected as the warfighting commander of Joint Task Force (JTF) South. The overall commander of the operation was Gen. Maxwell Thurman, commander of the US Southern Command (SOUTHCOM). The official notification of an impending invasion occurred at 1600 on December 17 with a call from the Joint Special Operations Commander, Maj. Gen. Wayne A. Downing, a plank holder in 1974 with the 1st Battalion, the first commander of the 3rd Battalion in 1984 and the first Regimental Commander in 1986. The Regimental Commander at the time was then-Col. William F. "Buck" Kernan who eventually would go on to retire at the rank of General and in command of Joint Forces Command (JFCOM).

The alert was passed to each Battalion. Lt. Col. Robert Wagner commanded the 1st Battalion and would retire at the rank of Lieutenant General. Lt. Col. Alan Maestas commanded the 2nd Battalion and Lt. Col. Joe Hunt commanded the 3rd Battalion.

Al Dochnal, now a retired Colonel, commanded C Company, 3rd Battalion; the company that was assigned to 1st Battalion for the mission. Dochnal and his commander Lt. Col. Hunt were briefed that night on the mission. Dochnal reported with his company to 1st Battalion the next day. The 1st Battalion leadership at Hunter Army Airfield tried their very best to keep the purpose of the recall a secret indicating it was just another readiness exercise testing the recall standards for a no-notice deployment.

That secret was not kept under wraps for long. But keeping the truth behind the recall was the least of the Rangers worries. By late afternoon the weather turned poor as a winter storm hit the southeast and the rain, sleet and wet snow made the working conditions miserable. The Rangers continued to toil as they worked in the open in the freezing temperatures – a stark contrast to the 90 plus weather and 100% humidity expected once they invaded Panama.

Regardless of the challenging weather conditions, the Rangers boarded the aircraft rigged to jump and departed with 16 aircraft full of Rangers at 1900 on the night of the 19th. Seven C-141s transported 1st Battalion while four C-130s carried C Company, 3rd Battalion. The remaining five C141s carried vehicles, supplies and equipment.

Seven hours later, at 0100 on December 20, preparatory fires preceded the 0103 combat jump when 732 Rangers from the 1st Ranger Battalion Task Force exited their aircraft 500 feet AGL over the objective.[59] Kernan, who led the airborne assault, stated that his Rangers had encountered hostile fire while still in the air, but the opposition, while "fierce at first," had "quickly died down."[60]

1st Battalion targets consisted of Objective Tiger which was the Fuerza Aérea Panamena (FAP); the Panamanian Air Force barracks; Objective Pig, the barracks of the 2nd Infantry Company in the center of the airfield; the Tocumen control tower; and finally Objective Bear, the main airport terminal assigned to the attached C Company, 3rd Ranger Battalion. In addition, other company missions consisted of securing the perimeter of Condor; preparing the runway for follow-on air landings; and securing Objective Hawk. Hawk was the Ceremi Recreation Center in the La Siesta Military Resort Hotel and was potentially a hideaway for Noriega and/or his Dignity Battalions.

The seizure of Tocumen went smoothly. With the southern security established, A Company quickly overwhelmed the FAP personnel who elected to fight and secured the barracks and their nearby aircraft. Part of C Company's objective, the FAP's 2nd Rifle Company's barracks, had been leveled and completely destroyed by a Spectre gunship allowing the Rangers to quickly secure the barracks. The second phase of their mission; the securing of the control tower, met stiffer resistance. Nonetheless, by 0210, twenty-five minutes later than planned, the 1st Ranger Battalion's objectives on the Tocumen military airfield were cleared and secured.[61]

South of the 1st Ranger Battalion; C Company, 3rd Ranger Battalion was involved with stiffer resistance in their effort to secure a fire station, an airline baggage area, and establishing an overwatch position prior to assaulting the terminal. With the fire station secured, 3rd Platoon, C Company continued on to the main terminal. Shots rang out from the northern rotunda, shattering glass, as the platoon moved across the tarmac. After a brief reprieve of gunfire, the Rangers began to sweep through terminal's ground floor. Following a brief standoff with five PDF soldier's holding some civilian hostages, the terminal was secured by 0500. By dawn, all of the Ranger objectives on Tocumen and Torrijos had been cleared and secured.

Dawn of the 20th found one major objective yet to be achieved; the PDF headquarters at *La Comandancia* in Panama City. The task fell to C Company, 3rd Ranger Battalion. Following preparatory fires and supporting building clearing assaults that commenced at 1500, the Rangers swept through *La Comandancia* and secured it by 1700.

By now, the 2nd Ranger Battalion linked up with the 3rd Ranger Battalion, minus C Company that had already departed early that morning to support the 1st Ranger Battalion. Both battalions constituted Task Force Red-Romeo, whose objective was Rio Hato, home base to the 6th and 7th PDF Rifle Companies. These companies were considered the best fighting units in the PDF.

Rio Hato was located approximately sixty miles southwest of Panama City on the coast of the Gulf of Panama what was designated "Area of Operations Eagle," or AO Eagle. The objective was divided into two different operational regions for each attacking Ranger battalion. The southern portion was assigned to the 2nd Ranger Battalion and included the 6th and 7th Company compounds; Objectives Cat and Lion. The two companies of the 3rd Ranger Battalion were to secure the northern sector of Eagle to include the NCO academy; the camp headquarters; the airfield operations complex; the motor pools; the communications center; and the ammunition supply point at the far northern end of the runway. Additionally, the battalion was to clear the runway for air landing operations to follow and cut the Pan American Highway that ran through the area.

The attack on Rio Hato commenced simultaneously with the assault on Torrijos- Tocumen. Three minutes after preliminary fires commenced, the green light of the lead C-130 lit and 837 Rangers exited the blackened aircraft at 500 feet AGL. The antiaircraft fires were heavy and, for the moment, unsuppressed as close air support had to cease-fire and withdraw while the paratroopers were in the air. Eleven of the thirteen pax transports were hit as they over flew the objective.

On the ground, the Rangers found themselves in a 360-degree firefight against an alert and dispersed PDF who, while not well organized, were everywhere, firing at the Rangers as they discarded their chutes. Despite the enemy fires, the Rangers fanned out to secure their objectives. Seven minutes ahead of schedule, one hour and fifty-three minutes into the operation, the airfield was reported secure.

At 1000 on December 25, the 3rd Ranger Battalion air-assaulted into the town of David to secure the Malek Airfield. Within the hour, the attack on the small airfield was completed and the facility seized.

Task Force Red-Romeo's final mission occurred on December 28 when two companies of 3rd Ranger Battalion assaulted Camp Machete, a penal colony on Coiba Island. The Rangers arrived to secure all the prisoners who had broken free from their locked cells. Hostile guards were rounded up and flown off the island. The next day, the Rangers were relieved by elements of the 7th Infantry Division.[62]

Operation Just Cause was by all measures a success for the newly formed Regiment. The Rangers captured 1,014 enemy prisoners and seized over 18,000 arms of various types by the time the operation was over. All in all, the heaviest concentration of casualties for the invading US force came from the US Army Rangers one in 18 Rangers were killed or wounded during their nighttime parachute assault and subsequent attack on Noriega's forces.[63]

STORMING THE DESERT

On August 2, 1990 Saddam Hussein, the tyrannical leader in Iraq, stormed into the oil rich country of Kuwait and laid claim to the tiny nation. Six days later, elements of the 82nd Airborne Division, most notably the 325th Airborne Infantry Regiment (AIR) from the 82nd Airborne Division, ironically under the command of then-Col. Ron Rokosz, the first commander of B Company, 1st Ranger Battalion deployed under a no notice deployment to Saudi Arabia to posture in a defense mode to protect further Iraqi aggression.[64]

For eight months President Bush ordered a massive buildup of forces in the Saudi Kingdom. His intent was to thwart the Iraqi push south and to posture the forty plus nation coalition built by the US so that it could remove Hussein's forces from Kuwait. The Ranger involvement in this conventional array of forces and intended high intensity conflict was small.

The commanding general during the Gulf War, Norman Schwarzkopf Jr., was reportedly not very fond of Special Operations Forces (SOF) in general. Though it is a hard pill to swallow, the general is correct in his assertion that SOF units do not by themselves win wars and this clearly was a conventional war. Nonetheless, the Ranger Regiment did inject itself in a meaningful way into the Desert Storm fight.

Regardless of the challenges to move to the sound of the guns in the first Gulf War, the Regimental Commander, Wayne Downing deployed elements of Company B and 1st Platoon Company A, 1st Battalion, 75th Ranger Regiment to Saudi Arabia from February 12, 1991 to April 15, 1991. B Company was commanded by then-Capt. Kurt Fuller (now serving as a special operations General Officer). The staff element from the battalion was led by Ken Keen, a retired Lieutenant General and former 75th Ranger Regimental commander.[65]

The Rangers were given critical tasks to secure and occupy objective rally points, conduct leader's reconnaissance of clandestine objects, and conduct raids. Established initially as a quick reaction force in cooperation with Allied forces; there were no Ranger casualties. It was reported by many that the Rangers actually performed a combat parachute jump in Desert Storm but no combat jump stars were awarded.

While their limited involvement didn't please the Rangers, the performance of these Rangers significantly contributed to the overall success of the operation, and upheld the proud traditions of the past. Following the Rangers involvement in Desert Storm, albeit limited, the Chief of Staff of the Army, Gen. Gordon R. Sullivan, would develop his own charter for the Ranger Regiment that included, "The Regiment must remain capable of fighting anytime, anywhere, against any enemy, and winning."[66]

SOMALIA – TASK FORCE RANGER IN GOTHIC SERPENT

Just over two years after Desert Storm, Company B, 3rd Ranger Battalion deployed to Somalia in August 1993. The mission in Somalia was to assist UN humanitarian forces as part of Operation Restore Hope. That mission evolved into a more aggressive task to reduce the threat to the UN mission. On October 3, 1993 the Rangers conducted Gothic Serpent with Delta operatives with the task to capture two of warlord Mohamed Farrah Aidid's lieutenants. The ensuing 18 hour gun fight with Somali guerrillas became widely known as the fiercest ground combat for US military personnel since the Vietnam War.

The situation in Somalia in 1993 was grounded in a volatile and violent history. In 1991 its government collapsed after suffering years of civil strife. The Somalian infrastructure was deteriorating and made worse by a drought that claimed thousands of lives Starving Somalis littered the streets and food relief missions initiated by NGOs and other private organizations were challenged in providing help. Warlord led clans confiscated what international relief that was provided to fuel their own military arms in an effort to control their turf.

Somalia had approximately 14 major factions with a litany of clans participating in domestic warfare. These violent clans hampered relief efforts to such a point that the United Nations approved Resolution 751 in April 1992 providing humanitarian aid to Somalia (UNOSOM I). UNOSOM I was unable to accomplish the mission properly due to a lack of resources. As a result the United States initiated Operation Provide Relief, a mission that added additional security measures. This operation sought to provide military as well as humanitarian support to the war torn country.

In March 1993, Operation Restore Hope was turned over to UN Peacekeeping Forces creating UNOSOM II, making it the first ever UN-directed peacekeeping mission. The UN mission had three distinct phases: disarm Somali clans, rebuild the political infrastructure, and create a more secure environment. Twenty-one nations contributed to UNOSOM II. The UN Secretary General appointed Turkish Gen. Cevik Bir as the commander of the force and selected retired Admiral Jonathan Howe as the special representative to the UN Secretary.[67]

A number of local clans were threatened by the nation-building effort. Most notably was the Mogadishu based Somali National Alliance (SNA) and its leader, Mohammed Aidid. Aidid was a former senior Somalian diplomat and military Chief of Staff when Maj. Gen. Mohamed Siad Barre staged a bloodless coup d'état resulting in the imprisonment of Aidid.

Trained in guerrilla war, Aidid quickly initiated a campaign against the UN which ultimately led to a series of ambushes killing 24 Pakistani peacekeepers. These soldiers were brutally torn apart and dismembered by the drugged-up clansmen. The UN had to demonstrate its willingness to protect its peacekeepers and the next day the UN and the White House responded. Under the leadership of UN Secretary General Boutros-Ghali and US Secretary of State Madeline Albright, Resolution 837 was passed authorizing action against those held responsible, namely Aidid. Strained relations between the Clinton administration and the military led to the exclusion of any input from the Pentagon regarding Resolution 837. This lack of confidence in the military would have a severe impact in the very near future.

The humanitarian mission was now a military one. Boutros-Ghali was a longtime enemy of Aidid, and was unable to resolve the escalating military situation. A $25,000 bounty was place on Aidid and that amount increased as the difficulties with Aidid and his clan continued. The battles became increasingly violent and personal. At this point military support was requested by all involved. To answer the calling was the SFOD-Delta and US Army Rangers.[68]

TASK FORCE RANGER

The Ranger force, known as Task Force Ranger (TFR), was under the command of Maj. Gen. William Garrison. TFR went after Aidid and his aides with gusto and, although numerous raids took place, Aidid was never caught. The support this effort, additional firepower in the form of mechanized vehicles and armor, was requested. Incredulously, the request was denied by the Secretary of Defense Les Aspin.[69] The Clinton-Military strain was having an impact.

In an irresponsibly poor decision, the White House, secretly opened negotiations with Aidid. At the same time the pressure on the US force by the administration increased. Frustrated and under undo pressure from the same administration that showed a lack of support, TFR undertook an unnecessary but daring daylight raid, resulting in the battles of October 3–4, 1993. It was here that 18 Americans died and Somali casualties ranged from 350–700 killed and thousands wounded. The White House withdrew US troops shortly thereafter.[70]

TFR consisted of Rangers from Company B, a platoon from A Company and a command and control element of 3rd Battalion, 75th Ranger Regiment. This force was deployed to Somalia to assist United Nations forces in their mission from August 26, 1993, to October 21, 1993.[71] The tasks in the UN Resolution included the apprehension of those responsible, namely Aidid, for the ambush and death of twenty-four Pakistani UN peacekeepers. The use of "all necessary measures" was authorized by the UN.[72]

The June 17 UN issued arrest order caused Aidid to go into hiding and efforts by in-country security teams to find him failed. As a result, Howe requested 1st Special Operational Detachment-Delta to assist in Aidid's capture. Clinton eventually approved the request and sent in the specialized unit.[73] The 450-man task force consisted of sixty men from the one-hundred-and-fifty-man C Squadron, 1st Special Forces Operational Detachment-Delta: B Company (Reinforced), 3rd Ranger Battalion, 75th Ranger Regiment; and support helicopters from 1st Battalion, 160th Special Operations Aviation Regiment (SOAR) – the "Night Stalkers."[74]

Task Force Rangers' advance party arrived in Somalia on August 26. They set up base on the shore of the Indian Ocean at the Mogadishu Airport on the far southern end of the city. The operation was to be conducted in three phases: Phase I, lasting until August 30, was to establish a base camp; Phase II, lasting until September 7, focused exclusively on locating and capturing Aidid; and Phase III, in the event Phase II failed, the focus would shift to Aidid's command structure with the intent of forcing the warlord to take a more active and open role with his forces.[75]

In spite of a number of handicaps, Task Force Ranger attempted to seize and to maintain the initiative by planning and launching a number of raids that proved unsuccessful. On September 7, the force moved to Phase III and expanded its target list to include six of Aidid's top lieutenants and staff. Despite marginal Ranger success, Aidid continued remained in hiding and managed to thwart a number of Ranger offensive operations that were executed leading to the final set on October 3rd and 4th.[76]

OCTOBER 3-4, 1993: MOST INTENSE FIGHTING SINCE VIETNAM[77]

Task Force Ranger's seventh and final mission commenced at approximately 1300 on October 3. Getting a tip from a Somali agent that a number of Aidid's lieutenants, including two of the six on the expanded target list, would be meeting later that afternoon, the assault force consisting of seventy-five Rangers and forty Delta soldiers onboard an air armada of sixteen helicopters departed for their target.[78]

The target was in the vicinity of the Olympic Hotel, a white five-story building that was one of the few large buildings left intact in the city. Cross sectioned by a number of narrow dirt alleys, Hawlwadig Road ran in front of the hotel and was one of the few paved roads in the city. One block north of the road was the target house, a two-sectioned building with two stories in the front, three stories in the rear, and a flat roof on both. L-shaped, the structure had a small courtyard enclosed by a high stonewall.[79]

Just three blocks to the west of the hotel was the Bakara Market. This market was known to be the most heavily armed region of Mogadishu. It was known by soldiers as "the Black Sea" and was referred to as real "Indian country."[80] The Delta operatives and supporting Rangers would be inserted by fast rope from four MH-6 Little Bird aircraft and six MH-60 Black Hawks transports with four AH-6J Little Birds providing close air support.

Because the target areas was too confining to land helicopters to extract the prisoners and assault force, a fifty-two-man Ranger ground element deployed from the airport in a twelve-vehicle convoy to eventually link up with the rescue force that would fast rope onto the objective. It was a three-mile movement to the objective area. After being delayed for 37 minutes, the helicopters departed at 1532.

Moving low and fast over the ocean's breakers along a circuitous route, the aircraft made a dash over the city, with the MH-6s carrying four Deltas perched on the skids. Landing quickly on Hawlwadig Road forty operatives exited the aircraft while sixty Rangers fast roped in to secure the perimeter.[81]

The "precious cargo," or PC, consisting of twenty-four prisoners, including the two primary men they had sought, were quickly captured. 200 meters from the objective the Ranger twelve-vehicle ground convoy mustered at a rally point awaiting the call to extract the PC. During movement to the extraction point the ground convoy took intense fire. As the ground convoy picked up the prisoners in front of the building, enemy fire began to gain in intensity.[82]

The situation grew progressively worse when a Rocket Propelled Grenades (RPG) was fired at US helicopters circling overhead. At 1620, the tail rotor of the lead assault Black Hawk, *Super 61* was hit causing it to crash on the roof of a house located within a walled compound. With its nose crashing into the ground the pilot was killed and five others were injured. Proficient at downed-aircraft-recovery-operations, the task force quickly implemented three contingency plans: provide cover with a nearby Combat Search and Rescue (CSAR) Black Hawk, *Super 68*; deploy the TFR main body from the objective to the crash site; and alert the 10th Mountain Division Quick Reaction Force (QRF) to deploy from its location at the Somali National University to the Mogadishu Airport, where it could launch to support CSAR missions.

The Rangers and Delta operatives at the objective area began to move to the downed Black Hawk location once the transfer of prisoners was complete. At the crash site, survivors were desperately establishing a defense while a Little Bird audaciously set down in a nearby alley called Freedom Road to extract the two survivors.[83]

Two Ranger platoon leaders, Lt.'s Tom DiTomasso and Larry Perino, quickly moved their Rangers east, trading fire with Somali gunmen also in a foot race to the crash on parallel streets. In the meantime, the fully loaded CSAR aircraft took an RPG round and limped back to the airfield before crash landing. At this point the 90 or so Rangers consolidated at the downed aircraft were caught in a killing zone of small arms fire and multiple RPG engagements. Within an hour, 10 of the 13 men with Perino would be wounded and Delta and the search-and-rescue team suffered comparable casualties.[84]

At the crash site it became apparent that the pilot killed in the crash, CWO Clifton P. Wolcott, was trapped in his seat. Braving hand grenades, rocket-propelled grenades and rifle fire, a half-dozen soldiers tugged and futilely power sawed in vain at the crumpled wreckage trying to free the dead comrade.[85] Abandoning their aviation comrade was not an option for the special operators.

At this point things got worse. Orbiting overhead fulfilling *Super 61s* operational space before being shot down was *Super 64*. While the task force struggled to recover their dead, *Super 64* became the next aircraft casualty taking a direct hit from an RPG round to the tail causing it to crash.

Super 64 crashed 1500 meters southwest of *Super 61*. One aircraft down was not good, but TFR's contingency plans could cover such an event. A second Black Hawk down, however, was never seriously considered and the distance between the two added to the complexity. The only forces available were already committed to battle and the only search-and-rescue team had already fast-roped down to Wolcott's aircraft. Therefore, there was no easy way to reinforce a second crash site. Nobody had taken seriously the prospect of *two* helicopters going down.[86]

Instead of back hauling the enemy prisoners by wheeled convoy to the airfield, the Ranger battalion commander, Lt. Col. Danny McKnight, was ordered to reinforce the crash site. Taking on this mission, McKnight's ground force had difficulty navigating the confusing labyrinthine of streets and alleys. To make matters worse, gunfire raked the eight vehicle convoy constantly along the route.

High overhead, the command-and-control aircraft attempted to direct the convoy's movement. Even though the convoy vehicles were marked atop with large fluorescent-orange panels (VS-17 panels) rising casualty count reluctantly forced the convoy back to the airfield. Hours earlier, requests to insert Delta snipers on board Little Bird aircraft was denied. On its third request the command was desperate and authorized the insertion of MSG Gary Gordon and SFC Class Randy Shughart onto the crash site. The command made this decision considering the reaction force convoy, upon which the command's hopes had rested, had been forced to turn back.

The intent of the insertion was to have the two men, each armed with a sniper rifle and a pistol, provide first aid, establish a defensive perimeter, and secure the site until the arrival of a rescue force. It was apparent no rescue force would arrive in time before the growing number of enemy closing in on the crash site would overwhelm them. All concerned knew the eventual fate for the two Delta NCOs once on the streets of Mogadishu.[87]

But Gordon and Shughart also knew that the four wounded men below would not survive without at least an attempt to help. Gordon's widow, Carmen, described her husband's actions later at a ceremony. "Gary went back to save his fellow soldiers, not to die there. Gary was one hundred percent Ranger. He lived the Rangers' creed every day. He knew that he had a chance. He and Shughart wouldn't ever have gone out there trying to be heroes."[88]

Attempts to insert the operatives were plagued by confusion caused by the chaos of the battlefield. Disoriented and receiving guidance from *Super 62,* in the form of hand and arms signals and smoke grenade emplacement to help vector the two men to the crash site, Shughart and Gordon were last seen signaling thumbs up as they fought their way, "under intense small arms fire through a dense maze of shanties and shacks to the downed Black Hawk."[89]

In the wreckage of *Super 64,* all four crewmen had survived the crash. The pilot, CWO3 Michael Durant was knocked out by the impact. When he regained consciousness he found the femur of his right leg broken and a large sheet of tin punched through his shattered windshield and draped over him. The copilot Ray Frank had his left tibia broken. Both pilots had sustained back injuries. Unable to move, Durant secured his German MP-5K 9-mm rifle and prepared to defend himself from his seat as the copilot crawled from the wreckage out the opposite side.[90]

Durant was relieved to see the arrival of Gordon and Shughart. The wounded pilot was extracted from the downed aircraft carried to a nearby tree next to the mortally wounded SSG William Cleveland who laid unconscious and bleeding. Cleveland was eventually defiled and drug through the streets of Mogadishu by jeering Somalis. Gordon and Shughart moved to the left side of the chopper to extract the remaining crew chief, SSG Thomas J. Field. Field survived the crash, but was killed by the attack mob when he joined the two Delta sergeants engaging the approaching militia and defending the exposed side of the downed helicopter.

Following Durant's brief explanation of procedure, Gordon established radio contact, requesting immediate help. The reply, as it had been before his insertion, was that a reaction force was en route to their location. With that, the Delta sniper gathered his weapons and moved back around to the left side of the aircraft to engage the advancing militia.

Out of ammunition, Gordon returned once again to the wreckage, looking for anything to fight with, only to find very little. Gordon handed a loaded CAR-15 automatic rifle to Durant, whose own 9-mm weapon was either out of ammunition or jammed. Telling him "Good luck," Gordon made his way back to the far side armed only with a pistol.[91]

Gordon was killed leaving Durant to futilely defend himself. The crowd had already lifted the lifeless bodies of his comrades to parade them through the streets and Durant fully expected to die at this point. Circling above, surveillance helicopters recorded images of "indigenous personnel moving around all over the crash site."[92] Nearly two hours after the aircraft had gone down the battle for *Super 64* was over. Everyone was dead except for Durant who was spared, but not after being held hostage for eleven days.

With the air coverage gone, Gordon and Shughart were on their own facing an onslaught of Somali militia advancing on the crash site. They defended the site valiantly but they both perished leaving Durant alone. Shughart was killed first and Gordon was left attempting to call for support unveiling for Durant that the two Delta men had arrived at the site on their own, with no other support

A Quick Reaction Force (QRF) of 425 men led by 2nd Battalion, 14th Infantry Regiment, 10th Mountain Division and augmented by Pakistani M-48 Tanks and Malaysian Armored Personnel Carriers finally deployed to support the fight, but obviously too late. Being ambushed all along the way the convoy eventually made it to both crash sites in the early morning on October 4 to link up with the Rangers at *Super 61* and to find no survivors or bodies at *Super 64*.[93] The fiercest ground combat for American forces since Vietnam was now over leaving a mark on the history of the modern day Ranger and setting the bar high for audacious and courageous performance by Ranger units in the future.[94]

HAITI; UPHOLDING DEMOCRACY WITHOUT A FIGHT

Regardless, of the string of failures by the National Command Authority and its employment of forces in Somalia, the Rangers had made a name for themselves as a joint force in that fight. So when the word came for military might in the Caribbean nation of Haiti, the powers to be turned to the 75th.

The 1st and 2nd Battalions and a Company of the 3rd Battalion, 75th Ranger Regiment were called out for a no notice deployment to Haiti for Operation Uphold Democracy in 1994. The operation was canceled on September 18 within five miles of its execution when a team of negotiators, dispatched by President Bill Clinton and led by former President Jimmy Carter, was able to convince General Raoul Cédras to relinquish power. The rigged for jump Rangers and those flying in on helicopters in from the carrier to fast rope had to abort.

Elements of the 1st and 2nd Battalions did however operate in-country while order was being restored. This is also the first operation where the US Army was the primary operating force on a US aircraft carrier, the USS *America* (CV-66).[95] The ship had Special Operations Forces from USSOCOM composed of Rangers, Special Forces, and other special warfare groups. On board that carrier was Specialist Ed Caraccilo, US Army Ranger, the author's brother and a Carl Gustav gunner in 1st Ranger Battalion.

THE YEARS LEADING TO THE WAR ON TERROR

The Regiment has evolved greatly since its foray in Somalia. Keith Antonia, a former Regimental Operations Officer, describes his command of 186 men in a company in the early 1980s. That command organized effectively for the times as the Rangers evolved it has changed greatly since. Today and in the years leading to Operation Enduring Freedom (OEF) and Operation Iraqi Freedom (OIF) the Ranger units have grown substantially including Stryker vehicles organic to the command. The Ranger Reconnaissance Detachment (RRD) was formed in the 1980s and grew to a formidable company size element during the War on Terror.[96]

In the late 1990s, Stan McChrystal, the Regimental Commander in the late-90s who would go on to retire as a 4-star general and commander of forces in Afghanistan, proved a visionary. His vision of how the Rangers should be organized and how they should fight was fortuitous for the requirements in the War on Terror. During his tenure he implemented the mortar arms room concept giving the Rangers the ability to choose from 60mm, 81mm or 120mm mortars depending on the mission set.

Additionally the Rangers arguably were the first Army unit to develop a web based command and control system where all planning and execution could be done via the intranet driven Ranger Web. The efficiency of this paperless system was evident in the Ranger Tactical Operations Center (RTOC) where those Rangers that ran the fight sat quietly listening to it through headphones and executing operations through a tethered command and control computer system.

One of the critical changes that occurred during the McChrystal tenure and followed up by his successor then-Colonel Ken Keen was the development of the Ranger Support Element (RSE). One of the critical weaknesses recognized by combatant commanders in the past few decades was that the Rangers were a light unit with no logistical capabilities. While the RSE was not organic to the Regiment, the riggers, maintenance staffs, truck drives and other logistics elements were dedicated to the Regiment and the criticality in out loading them for a fight. Being integrated totally into the JSOC command provided the Regiment those logistics assets required once they deployed.

Not only had the organizational structure evolved, the training regimen has changed as well. The inclusion of Ranger Medical Training so that each Ranger could keep his "buddy" alive in combat, a focus of hand to hand combat, intense foot marching, night firing, small unit battle drills and other anti-terrorist focused efforts made the Regiment completely prepared for the decade long combat eras in the Middle East yet to come.

The years after Somalia leading to the Global War on Terror saw sporadic Ranger involvement in the small skirmished that involved US combat power. On November 24, 2000 the 75th Ranger Regiment deployed the Regimental Reconnaissance Detachment (RRD) Team 2 and a command and control element to Kosovo in support of Task Force Falcon. As the years unfolded a number of terrorist attacks like the one on the USS *Cole* off the coast in Yemen on October 12, 2000 helped focused the Leuer Performance Training concept to that of an anti-terrorist flavor. Regimental Commanders like Stan McChrystal and Ken Keen, placed great emphasis in the late 90s and early 2000s on the establishment of the US Ranger Regiment as a Task Force (Red) within the Special Operations community, most notably JSOC, so that when the towers were hit on 9/11 the nations go-to force was ready.

6

Forging a Special Operations Force – Rangers in the War on Terror

Valor is stability, not of legs and arms, but of courage and the soul.
                    -Michel de Montaigne, writer during the French Renaissance

When the 9/11 terrorist attacks on the US occurred the Ranger Regiment was already a formidable special operations force. This is also the time where Abrams' Charter truly became a reality as the rest of the Army benefited from its postulating concepts.

As the 1990s unfolded the 75th Ranger Regiment became a critical part of the special operations community within the Joint Special Operations command, all the while presenting itself as the best light infantry force in the world. Night operations, foreign deployments and a disciplined integration of life fire exercises continued to be a Ranger training staple. Yet, even though they had proven their worth in battle, many special operations missions seemed to rely on the Regiment to be outer ring for tier 1 units like Delta. This attitude would change with the onset of the wars in the Middle East.

There is no doubt that the rest of the US Army desired to operate like the Rangers. Moreover, in order to survive in what would become a fast past counterterrorist operation in the Global War on Terror (GWOT) in the Middle East, the Big Army, out of necessity, had to adopt the Ranger way of training and executing operations as the war unfolded. The special operations community operated on a very definitive set of principles: simplicity, security, repetition, surprise, speed and purpose and these principles would find their way into the core of the conventional army as time unfolded.[1] As the wars in Iraq and Afghanistan unfolded – the special operations and conventional lines would blend together causing the conventional forces to instill these principles in their daily operations in order to achieve victory on the battlefield.

One would argue that the Global War on Terror (GWOT) was the ultimate time to be assigned to a Ranger battalion. Decades of Ranger experiences and sacrifices in smaller skirmishes had polished the Ranger standard operating procedures. These SOPs would prove useful allow the Rangers to excel in the counterinsurgency fights in Afghanistan and Iraq.[2]

As the war continued the Army was becoming more and more like the Regiment. Notably, the way the Regiment gathers information on the enemy using special equipment both in signal intelligence and other means was proliferated to the rest of the Army. Additionally, the methods the Rangers used to form their plans and conduct their operations, particularly raids and small unit attacks, was emulated by the conventional forces. While the Big Army tended to imitate the operational proficiency of the Ranger units, this emulation was, in reality, done out of necessity given the operational environment in OEF and OIF.

There were visible changes as well. As stated earlier, the Rangers wore the distinctive black beret. The color black signified color of night operations and indicative of the time of day the Rangers operated. As the Rangers became more and more successful and respected, the black beret quickly became a sign of elitism. The rest of the Army looked to the wearer – the Ranger – as one who was the best of the best and wanted to be like him.

In 2001 the Chief of Staff of the Army, Gen. Eric Shinseki, ordered the adoption of the black beret for the entire US Army. He had looked to the respect the world had for the Ranger black beret and knew that by fitting the rest of the army with this distinctive, proud headgear that the rest of the Army would stand as proud.[3]

The 21st Century Ranger Regimental Scroll. (Source: Caraccilo)

The 21st Century 1st Ranger Battalion.

Rangers, having taken umbrage at the alleged insult regarding their black berets, should have viewed it as a compliment instead and realized the importance of the ongoing struggle within the Pentagon. The Regimental Commander at the time, Ken Keen, made this statement regarding the Army's decision to adopt the black beret and the Regiment's decision to adopt a new colored beret in response: "Unity within our Army is absolutely critical to combat readiness and Rangers have always prided themselves in being part of that unity. Unity among Rangers, past and present, is essential to moving forward and ensuring we honor those who have put the combat streamers on our colors and acknowledge the sacrifices and dedication of the Rangers and their families who serve our nation today."[4] In June 2001, some three months before the attack of 9/11 the Rangers removed their black berets forever and donned the new identifiable Ranger headgear; the tan beret.

TERROR STRIKES

On 9/11 a New World Order had begun. Between 0845 and 1010 on that fateful day in September 2001 four American commercial airliners hijacked by 19 suicidal fundamentalists brought havoc on four different targeted areas in the US. The terrorist flew two jumbo aircraft into the twin towers at the World Trade Center, one at the Pentagon and the last one crashed in a Pennsylvanian field. A total of 2,977 people were killed in New York City, Washington, DC and outside of Shanksville, Pennsylvania, in the worst terrorist attack in U.S. history.[5] Many have called 9/11 the second Day of Infamy relating it to the Roosevelt announcement in December 1941. Not since that infamous day had the nation reacted to an incident with such fervor. War was on the horizon.

President George W. Bush's initial reaction to the attacks was guarded. He was visiting a classroom when the attack occurred. The President's first emotion was of anger and disbelief, stating that it was 'like watching a silent movie' and quickly realized that a lot of people would be looking for his response telling reporters that the Nation's response to the terrorist attacks would be "measured" and "decisive." [6]

Immediately after the attacks he told the US public that he was determined to find the culprits. He was intent on searching for those "who were behind these evil acts" directing the full resources of the US intelligence and law-enforcement communities to find them and to bring them to justice. There was no doubt that the 9/11 attacks had changed the world forever.[7] The War on Terrorism was under way and the Rangers were the lead act.

When the attack happened, Gen. Tommy Franks, the Central Command Commander (CENTCOM) was on his way to Pakistan to meet with President Pervez Musharraf about security considerations and counterterror possibilities when the news of the tragedy made his way. He was diverted back to his command headquarters in Tampa to begin directing the eventual military response. The entire US Army went on alert that night. The 75th Ranger Regiment, of course, led the way.

While the National Command Authorities and their senior commanders huddled to develop a plan in reaction to the attacks, the author had just relinquished duties as the 75th Ranger Regiment Regimental Executive Officer to Lt. Col. John Castles. Duty as a battalion commander was calling and the author was prepared to stand up a new airborne battalion in Vicenza, Italy as part of the 173rd Airborne Brigade. However, there was a delay in movement to Italy due to diplomatic issues surrounding the establishment of an additional battalion in that country.

As a result it seemed prudent for the author to attend the pre-command course which he was initially forgoing in order to hurry up and get to Italy. He departed for Fort Leavenworth, Kansas on September 10, 2001. On day two of the course, while sitting in a computer simulation lab, the television set in the top corner of the room displayed the towers in New York falling. Immediately his beeper went off recalling the author to homebase at Fort Benning, Georgia.

Realizing that the nation's airspace was closed due to the terrorist attacks, the author immediately drove a rental car back to Fort Benning from Kansas. Arriving quickly as possible he went to the Regimental Headquarters where there was a bevy of activity. One officer, Maj. Sean Jenkins (a recently selected Brigadier General) was the Senior Liaison Officer for the Regiment and asked the author to watch his desk as he left to get married realizing that he would be deploying soon. While watching the desk he fielded a phone call from JSOC requesting a Lieutenant Colonel to immediately deploy to command the Forward Operating Bases for JSOC in its initial response to the 9/11 terrorist attacks.

Bringing this requirement to the Regimental Commander, then-Col. Joe Votel, Votel unceremoniously told him he was going to fulfill this duty immediately. Votel told him that he would be leaving with the JSOC Commander, then-Maj. Gen. Dell Dailey (retired Lieutenant General and US Ambassador at-large) who was in Fort Benning and enroute back to Fort Bragg, North Carolina in preparation for the imminent deployment. Days later the author was in Misirah, Oman establishing the Forward Operating base (FOB) as a staging area for the initial US attack into Afghanistan against Al-Qaeda operatives who claimed responsibility for the US terrorist attacks.

US RESPONSE TO 9/11

The towers in New York City fell on 9/11. That night the Rangers, as well as the rest of the US Military, was alerted. The 3rd Ranger Battalion was the designated as the "ready force" at the time and Fort Benning was alive with preparations to deploy to do something - somewhere. It was still unclear as to the mission but there was no doubt that a JSOC Task Force, including elements of the Ranger Regiment, was on a short string to conduct whatever mission the National Command Authority desired. Having deployed that fateful day following the 9/11 attack with Maj. Gen. Dailey, the author would learn about that mission and those that caused it firsthand.

Al-Qaeda ("The Base") was responsible for the attack. This terrorist organization was organized by the likes of Osama Bin Laden in the early 1980s to support the war effort in Afghanistan against the Soviets. [8] The victory over the Russians gave rise to the overall jihadist effort and an aggressive jihadist movement emanating from Afghanistan ensued. Finding its way to countries on the Levant and Saudi Arabia, Al-Qaeda began to refocus its efforts against the US and its allies

By 1989, Al-Qaeda dedicated itself to further opposing non-Islamic governments and began to impose their opposition with force and violence. The group grew out of the Maktab al-Khidamat (MAK), or the Services Office, organization, which maintained offices in various parts of the world.[9] At this point, Al-Qaeda began to provide training camps and support bases not only in Afghanistan but in Pakistan and the Sudan as well and began to set their sights on recruiting US citizens.[10]

Up until 1991 the terrorist group was headquartered in various locations in Afghanistan and Peshawar, Pakistan. In the early 1990s it relocated to the Sudan where it was headquartered until approximately 1996, when Osama Bin Laden, the terrorist cell's military chief Mohammed Atef and other members of Al-Qaeda, returned to Afghanistan. From the Sudan it was able to expand its network to various other parts of the world to expanding it footprint and increasing its monetary capability.[11] On February 22, 1998 Bin Laden directed all Muslims to kill Americans in the following message: "in compliance with God's order, we issue the following Fatwa to all Muslims: the ruling to kill the Americans and their allies, including civilians and military, is an individual duty for every Muslim who can do it in any country in which it is possible to do it."[12]

A series of Fatwas that followed directed various terrorist operations to include the starting of the logistical planning that led to the bombing of the US Embassies in Kenya and Tanzania. Al-Qaeda influence on the Taliban leaders in Afghanistan coupled with intelligence indicators that Bin Laden was mustering in that country motivated decision makers that a military operation into Afghanistan was the right choice for a visible US reaction to the 9/11 attacks. The mission quickly went to JSOC where plans were unfolding within hours of the attack.

Planning for the war in Afghanistan and the follow on fight in Iraq began in earnest.[13] As with most decisions to send American men and women into the fray, the habitual friction between the civilian and military leadership heightened. The Chief of Staff of the Army, Gen. Eric Shinseki testified to the US Senate Armed Services Committee that "something in the order of several hundred thousand soldiers" would probably be required for postwar Iraq. This was an estimate far higher than the figure being proposed by Secretary Rumsfeld in his invasion plans and would eventually cost Shinseki his job.[14] While the rumblings of how to project power and what power to project would be the centerpiece of media attention in the months following the 9/11 attack, a more focused approach to war-fighting within the special operations community was taking place in the weeks following the terrorist attack.

While there was definitely movement in the special operations community in the days immediately following the September 11 attacks, the public eye hadn't yet seen any overt signs of retaliation as a penance to what the terrorist did so blatantly to the US. The US leadership knew that Americans soon needed some sort of adequate response to the 9/11 attacks.

President Bush inextricably linked the attack on US soil to the Taliban who were harbored and training in Afghanistan. In the days after the 9/11 attack he gave Taliban leaders a series of clear and specific demands: Close terrorist training camps; hand over leaders of the Al-Qaeda network, and return all foreign nationals, including American citizens unjustly detained. None of this happened.

As a result, on October 7, 2001, President Bush declared: "On my orders, the United States military has begun strikes against Al-Qaeda terrorist training camps and military installations of the Taliban regime in Afghanistan. These carefully targeted actions are designed to disrupt the use of Afghanistan as a terrorist base of operations, and to attack the military capability of the Taliban regime... We did not ask for this mission, but we will fulfill it. The name of today's military operation is Enduring Freedom... Peace and freedom will prevail... May God continue to bless America."[15]

THE TIP OF THE SPEAR – THE US ARMY RANGERS

Having entered the JSOC compound at Fort Bragg the author would learn that the President directed the Central Command to conduct a forced-entry mission into Afghanistan for two purposes. The primary driver of the invasion was the September 11 attacks on the US, with the stated goal of dismantling the Al-Qaeda terrorist organization and ending its use of Afghanistan as a base. The United States also said that it would remove the Taliban regime from power and create a viable democratic state. Most importantly, as the author would learn early on, was that it was critical that the President show the American public that the US was reacting in some meaningful way to the 9/11 attacks.

JSOC formed a Task Force that would visibly project power into Afghanistan and display that power via the media to the American public. That was the initial task, but they first had to find a place from which to project that power. Planners started their process for defining the location of the Forward Operating Base (FOB), what forces where required and how the initial forced entry would be displayed to the American public.

What struck the author as a bit out of sorts was that there was no top driven mission objectives where locations and tasks, other than to get video immediately back to the Pentagon for dissemination, were directed by the National Command Authority. Planners from JSOC Headquarters, combined with operators from the Rangers, SEALs, Delta, the Night Stalkers, and others, to include the Special Operations Air Force community, mustered to define where to go and what to do. Locations like Uzbekistan, Oman, and various Gulf Cooperation Council (GCC) countries were all considered as a staging base and brainstormed for feasibility. Planners and leaders literally sat around the table calling out names of locations as possibilities for use and then a quick feasibility check was made to confirm or deny potential locations. In the end, JSOC decided upon the island of Misirah, Oman and an aircraft carrier, the *USS Kitty Hawk,* positioned in the Arabian Sea to launch its forces.

The objective chosen by the same group was a small airstrip named Operation Rhino deep in the Helmand Desert near Kandahar. Here the Rangers would conduct a low night time parachute assault to destroy the Taliban encampment located on the airfield. Immediately after the jump the task force would then flush the aircraft and off load little birds for Delta operators who would conduct an air assault raid on a compound in Kandahar to kill Taliban's Supreme Leader Mullah Omar. This first mission was a Ranger classic.

The Delta Force would then return to Rhino to exfiltrate back to Misirah with the rest of the Ranger Force. Supporting both operations was an air assault of B Company, 75th Rangers from the *Kitty Hawk* through a mission support site (MSS) codenamed Honda in Pakistan and into the objective area. There were no follow-on forces for this attack; instead it would be executed as a daring short nighttime raid.

In a unique exchange of ideas one afternoon before the jump Maj. Gen. Dailey asked a random group of operators who happened to be in the JSOC command center on Misirah what the Task Forces should be named. Serendipitously the name Sword was brought up and from then on out the Task Force would be formally known as Task Force Sword.

Misirah was a small hot and extremely humid location where the sterile Desert Storm era camouflage battle dress uniforms (BDU) became condensed with sweat and their porous nature was quickly consumed making their wear unbearable. The Air Force arm for the US Central Command was AFCENT and they were responsible for the life support on the expanding base on the island. Coupled with is commonly known as D-Cell, the JSOC logistics element, living in climate controlled tents was adequate for the preparation for the mission.

As the commander of the operating base the author was responsible for ensuring the security of the base was maintained and to coordinate all requirements through the Omani hosts. The author also had the fortunate experience of flying in the AC-130 fixed wing

Loading for the Combat Jump onto Objective Rhino.

aircraft, also known as Spooky which was an upgrade version of the Vietnam era Spectre, on the night before the jump. [16] The mission on this flight was to prepare the objective for the jump. This meant conducting fire missions on all known Taliban locations with the AC's 105mm artillery canon and 25 mm GAU-12 Equalizer. The crew allowed the author to fire the canon on the Taliban Tower destroying it on impact from thousands of feet in the sky and then destroying the Taliban terrorists as the fled the burning tower with the 25 mm gun.

On the night of October 19, 2001, a company-sized element of exactly 199 Rangers from the 3rd Battalion, 75th Ranger Regiment and the Regimental Headquarters departed from Misirah on four Lockheed MC-130 aircraft. The Rangers were JMPI'd (jumpmaster check of their serviceability of their parachutes before the jump) and loaded alongside motorcycles, little bird aircraft and other required equipment for the four hour flight to Rhino. The Regimental Commander, then-Col. Joe Votel directed that besides a five man element for command and control from the Regimental Headquarters, all jumpers would be "shooters." There was no room for fodder and Votel made that command directive very clear. There were no "straphangers." [17]

That night the author was fortunate to JMPI some of the jumpers and then depart on the P3 Orion, a surveillance aircraft that has the capacity to film the operation from 25,000 feet.[18] Once the filming of the jump was complete the aircraft saved the data and then made it available to quickly transfer to the JSOC command. From there JSOC forwarded images of the parachute assault into Afghanistan to the Pentagon for its public affairs mission of showing the US resolve by showing them that the US was not afraid to conduct operations in reaction to the 9/11 attacks.[19] This was a cutting edge method at the time for sharing data, which seems like an antiquated and primitive way of exchanging information. The world now has the ability to instantaneously share information with the simplicity of a keystroke. Nonetheless, the technology that existed was the best the community had and was used at the time.

Additionally, army photographers with night-vision cameras videotaped the parachute assault and the flushing and exfiltration of the Ranger force by MC-130 aircraft. Within hours of the jump, a combat camera quickly edited of the raid's footage and electronically delivered the package to a product development team of the 3rd Psychological Operations Battalion (POB) at Fort Bragg. Here the team edited and transmitted the video clip to the Pentagon in time for Secretary of Defense Donald Rumsfeld noon press conference on the following day.[20] Coupled with the video coverage from the P3 the National Command Authority's mission was accomplished.

In preparation for the Ranger's airborne drop several targets on and around the objective were serviced by US air power. Bombs dropped from B-2 stealth bombers, then by fire, as discussed, from orbiting AC-130 aircraft resulted in a number of enemy KIAs and several enemy fleeing the area. As with Ranger ritual the Ranger Creed was recited in each aircraft before the jump from the four MC-130 Combat Talon aircraft flying at low altitude over the objective in zero illumination.

The Ranger's objectives were to:

- seize the landing strip
- destroy any Taliban forces
- gather intelligence
- assess the suitability of the landing strip for future operations
- establish a forward aerial refuel/rearm point (FARP) for helicopters involved in the nearby operation in Kandahar called Objective Gecko"[21]

Typically in a combat jump the Ranger does a hop and pop leaving his expended parachute where it lands. However, with a flushing of fixed and rotary wing aircraft on the freshly secured airfield the need to recover and stow the parachutes was essential to ensure there were no hazards on the runways. A and C Companies of then-Lt. Col. Stephen Banach's battalion made the combat jump from the MC-130s that departed Misirah. B Company, 3rd Ranger Battalion flew on Task Force 160 Aircraft from the *Kitty Hawk* in the Arabian Gulf to support the mission. During their infiltration an MH-60 Black Hawk struggled with the brown-out conditions caused by dust and darkness and flipped over while attempting to land at the MSS Honda in Pakistan. Two Rangers were killed in the accident. SPC Jonn J. Edmunds of Cheyenne, Wyoming, and PFC Kristofor T. Stonesifer, of Missoula, Montana were the first casualties of Operation Enduring Freedom.

A Company and a team of snipers had the task to secure Objectives Tin and Iron and prevent enemy interdiction of Objective Cobalt and the landing strip. The Rangers were prepared for resistance, but only one enemy fighter engaged the approaching force. Several Rangers of C Company shot him. The mission for the C Company was to clear Objective Cobalt, the walled compound on Rhino. Although Spectre gunships had put devastating fire on the compound, damage was minimal. The thick rebar enforced concrete walls and building roofs had either absorbed the blasts or the cannon shells had punched through the roofs and ceilings leaving only holes. The guard towers were in the same condition.

A loudspeaker team from the 9th POB broadcast messages as the Rangers moved toward the walled compound, encouraging any remaining enemy fighters to surrender; but there was no one in the compound. The pre-assault fires from the AC-130 had taken its toll on the enemy force and willingness to defend their turf. The well trained Rangers continued to methodically clear the compound. Well-rehearsed teams quickly moved to clear their assigned buildings and guard towers. Clearing the interiors of buildings proved difficult and took longer than anticipated. A number of rooms inside had locked steel doors that could be opened only with multiple shotgun blasts or explosive charges With the landing strip secured, a MC-130 landed with medical personnel from JSOCs Joint Medical Augmentation Unit (JMAU) aircraft – basically a hospital on an airplane – and proceeded to treat two Rangers who had been injured during the jump.[22]

The ultimate goal, all Rangers knew, was to not only reduce the enemy capability on the airfield but to ensure the runway was capable of flushing the aircraft. To do this the US Air Force Combat Controller Teams (CCT) surveyed the landing strip, assessing it for possible future use. The CCT was also on the hook to communicate with the AC-130s circling high overhead. With the airfield cleared and secured, MH-60 and MH-47 helicopters, flown by the 160th SOAR and taking part in the operation at Objective Gecko, arrived and then refueled and rearmed at the FARP established using MC-130 tankers.[23] With all objectives completed, the Rangers and CCT boarded the MC-130s and exfiltrated back to Misirah. As they departed the force left PSYOP leaflets behind to let those that wander onto the camp that the wrath of the US is at hand. In the coming months, the US Marines used Rhino as a base camp for future operations.[24]

Later, thirty-two Rangers from B Company, 3rd Battalion made the second combat parachute assault on November 13, 2001. This hastily assembled mission was developed in a 24 hour time frame. The task was to establish a Desert Landing Strip (DLS) affectionately named Bastogne. This sight would be a forward arming and refueling point for future operations.

While the initial mission was highly successful there was much more work to be done in Afghanistan. What followed after the initial combat assault was 14 years of Ranger involvement in Task Force operations to kill or capture terrorists across the spectrum of operations in OEF. What made this task even more challenging was the commencement of Operation Iraqi Freedom (OIF) in February 2003. With two different theaters to operate, the Ranger involvement increased twofold.

FREEING IRAQ

As OEF unfolded the National Command Authority's attention turned to Iraq. It isn't within the scope of this work to discuss or analyze the validity of the decision to deploy forces to destroy the Saddam Hussein regime; rather the intent is to describe in brief snapshots the Ranger involvement in this daunting task. Other works have described in detail the high intensity conflict (HIC) involving the thunder runs to Baghdad by the likes of the vaunted 3rd Infantry Division (Mechanized) and the combat parachute assault by the 173rd Airborne Brigade on Bashur Drop Zone so it could link up with the 4th Mechanized Division in the north. It is also not the intention of this work to exhaustively define all missions the Rangers undertook during the 10 year war.[25] The intent for this work is, however, to show how the Ranger and the Ranger force has evolved through examples of excellence in combat.

The first few weeks of Operation Iraqi Freedom (OIF) were fast paced and complex. The US Military was attacking Baghdad from every cardinal direction to include from the air. Baghdad has been seized, Kirkuk and its prized oilfields were secured by Peshmerga and US forces and air superiority was clearly in the hands of the coalition. There was so many operations going on that the April 1, 2003 announcement about the seizure of the Haditha High Dam and Ramadi Highway Bridge went mostly unnoticed. Therefore, the taking of these two pieces of Iraqi infrastructure seemed a trivial thing. Not so.

The potential for the "scorched earth" tactics employed by Saddam Hussein during the 1991 Gulf War weighed heavy on coalition leadership minds. Environmental disasters via flooding of the sea and ports with crude oil, the destruction of desalinization plants, well-heads, gas-oil separation plants and physical substructure were just a few of the schemes the tyrannical leader used during past conflicts. It wasn't lost on the US and other members of the coalition that he may do it again. It would seem that the taking of two pieces of Iraqi infrastructure seemed a trivial thing. It was anything but.

Identifying key terrain is always an important part of any mission analysis in a military operation. In military terms, key terrain is any location where the seizure or retention of an area affords a marked advantage over the enemy. For the coalition the obvious key terrain included the oilfields in Basra and Kirkuk and the government footprint in Baghdad. Additionally, planners wanted to identify key infrastructure that would provide a marked advantage during the fight and to those in charge after the war. At the top of the list of bridges, power plants, and other structures was the Haditha High Dam over the Euphrates River.[26]

The second largest damn in Iraq, the Haditha Dam, at the time of the prewar assault in 2003, provided one-third of the Iraqi electricity. It is so important to the region that today extremist fighters from the Islamic State of Iraq and Syria (ISIS) are bearing down on it and if they seize it they could effectively limit the Euphrates' flow to the entirety of the Euphrates and Tigris River Valley and the flow of commerce into southern Iraq— much of which is Shi'ite.[27]

What concerned CENTCOM planners was that Saddam Hussein would order the locked gates open or destroy the dam with the goal of flooding the entire central part of the confluence of the Tigris and Euphrates' Rivers. Doing so, Saddam would essentially create an ecological and humanitarian disaster around the highly populated areas in the center part of the country.[28]

It was readily apparent to the CENTCOM planners that the Rangers and other SOF units were an integral part in the invasion of Iraq. Setting the mark high in Afghanistan in the two years prior to OIF, the Rangers had the credibility to seemingly accomplish any task assigned to them. As with the rest of the coalition in Iraq, the Rangers were initially set on finding and neutralizing weapons of mass destruction (WMD). In the form of missile warheads these weapons, fired from western Iraq, these missiles could easily attack targets in Israel and expanding the war in the region. To thwart this threat the coalition formed the Combined Joint Special Operations Task Force – West (CJSOTF-W) which was a mixed force of US Army Special Forces (SF), British and Australian Special Air Service (SAS), the 160th Special Operations Aviation Regiment (SOAR), and aircraft from the US Air Force (USAF), Royal Air Force, and Royal Australian Air Force (RAAF). To round out this force was the infusion of several battalion-sized infantry formations, including the 3rd Battalion of the 75th Ranger Regiment.[29]

Col. John Mulholland (an active duty Lieutenant General at the time of this writing) commanded CJSOTF-W. His mission was to conduct combat operations in western Iraq to locate and neutralize any tactical ballistic missiles (TBMs) armed with WMDs. To do this effectively Mulholland realized he had to seize number of airfields in the Anbar Province.

With little fanfare, Mulholland employed a force, including the Rangers, to seize a series of small airfields via parachute assault in an effort to find tactical ballistic missiles or WMDs. Jumps at both H2 and H3 proved to be "dry holes." Transitioning to a more reliable mission set the 3rd Ranger Battalion jumped again on All Qaim on the Syrian border on March 24, 2003. Here C Company, 3rd Battalion seized a small airfield allowing Company B to conduct an air assault on Night Stalkers' aircraft to raid the Al Qadisiyah Research Center named Objective Beaver, a suspected WMD testing facility three days later.[30] Finally, on March 28, Company A, 3rd Ranger Battalion reinforced with engineers from the 27th Engineer Battalion and a CCT element jumped from a C17 aircraft onto H1 Airfield named Objective Serpent, the northernmost of the Iraqi fighter bases. It took Mulholland's joint task force just 10 days to secure Al Anbar province, but no WMDs were found.

Mulholland moved the CJOTF-W headquarters to H1 and began to sorting out new targets for his command. Controlling the west from these key airfields, Mulholland could range several critical pieces of infrastructure. The first two targets were two of the most important pieces of infrastructure in all of Iraq: the Haditha Dam (Objective Lynx) and the Ramadi Highway Bridge. While it was obvious why the dam was important, including the Ramadi Highway Bridge to the mission set was just as critical. It was one of only two crossings on the Euphrates River capable of supporting heavy loads overland from Jordan; its destruction or interdiction would make supplying the cities of central Iraq all but impossible in the days following the fall of the Saddam Hussein regime.[31]

RANGERS IN ACTION

The Battle of Haditha Dam is a clear example of how the Rangers have evolved and set the clear path for how they would be used as a special operations force to be reckoned with in the future.[32] Seizing it by the evening of March 31, 2003, Company B, 3rd Ranger Battalion's task was to ensure that the dam was not prepared for destruction by enemy forces and to provide a line of communication across the river for follow on forces. Guarded by 100-200 enemy personnel, numerous armored vehicles including T-55 Battle tanks and in excess of 50 anti-aircraft artillery (AAA) pieces the defense appeared formidable. Because the Rangers were so proficient on combat planning and readying the force, orders and rehearsals were completed within four hours of conception of the mission. At the break of dawn, the Rangers departed on a 31 kilometer infiltration to the objective area.

About 2 km out from the objective, the Rangers recovered and fixed a disabled ground mobility vehicles (GMV) on the fly and began the assault with, surprisingly, no resistance. Arriving on site, the Rangers realized they underestimated the size of the dam and recognized this task was more daunting then initially thought. 3rd Platoon, Company C moved to the crest of the hill overlooking the entire dam where they realized the complexity of the mission. The top of the hill had 12 large concrete buildings the Rangers hadn't expected or developed a plan to attack. Immediately requesting more Rangers to clear the extra buildings they began the operation making sporadic contact.

Thus far the operation was without challenge until they moved to the far side of the hill to establish an overwatch position. All at once the lip of the hill opened up with small arms, machine gun and rocket-propelled grenade (RPG) fire. The Iraqis had moved out of the buildings to survivability positions at the base of the hill where they engaged the approaching Ranger force. Staving off the indirect fire from more 10 different mortar tubes, the platoon immediately returned fire and began what turned out to be a four-hour battle to push the enemy forces back at least 1,200 m out of RPG range.[33]

As the fight continued the situation progressively got worse for the Rangers. The vast, open desert in front of their defensive positions was a series of interconnected trench lines and bunkers for as far as the eye could see. As the day continued, periodic gunfights ensued with the enemy forces consolidating in 50 to 100 man sized groups six to eight kilometers from the Ranger location. The Rangers were able to successfully disperse the enemy consolidation plans, however, they would not wither away and continuously attack the Rangers in waves of 10-15 personnel.

This continued throughout the next few days. During the evening the Rangers began to fortify their positions and tracked the enemy movement with thermal image devices. Incredulously the enemy strolled up to the Ranger positions, arms slung over their shoulders as if they were on a walk in the park when the Rangers engaged them about 600 to 800 m out. This type of action continued for the next two days, with at least one of the Rangers positions in contact every 30 minutes.[34]

During the evening and throughout the second day, the Iraqi forces continued attempts to retake the dam. Around mid-day a kayak was seen heading toward one of the Ranger positions. With one burst from the .50 caliber, a Ranger gunner sank the kayak and the Iraqi began to swim to shore. The Rangers captured the uninjured Iraqi thinking he was conducting a reconnaissance mission on the Ranger force. After a brief search he was found to have a number of sketches of the Ranger positions.

Day three saw a series of indirect fire exchanges and at first light on fourth day, B Company's Battle Position (BP) began to be hit hard by artillery from what seemed to be every cardinal direction. Within minutes the artillery was impacting inside the perimeter and throughout the day artillery continued non-stop, moving from one BP to another. During this barrage, the Rangers received more than 100 artillery rounds within the perimeter and more than 350 on the entire dam. Later on day four, the Rangers experienced a number of close calls as the enemy became more proficient in their effort.[35]

On day five the Rangers didn't receive any artillery rounds. The continual barrage in the preceding days gave the Rangers an ability to reverse azimuth the artillery and mortar locations. Using aerial platforms, they now believed the aircraft had finally found the gun pieces and began firing on their location neutralizing them. Around mid-day on day 5, the Rangers were ordered to and clear CAS 1 and 2; the names the buildings gave to two of the buildings on the dam. Inside the buildings the Rangers found 12 rooms containing weapons and ammunition caches; it became clear to the Rangers now why the enemy kept moving to and from each. While the week would close out with more enemy artillery hitting sporadically on the Rangers, the CJSOTF force left the secure dam and continued operations to the north.[36]

The capture of the Haditha Dam is one of the most impressive examples of direct action by a special operations force during the war in Iraq and otherwise. Many relate the Ranger action on the dam to the seizure of Pointe du Hoc on the shores of Normandy in World War II. Regardless, as a testament to the Ranger perseverance to ensure the dam remained intact, the 75th Rangers was awarded the Valorous Unit Award for actions at Haditha. The operation was obviously valorous for individual Ranger awards for the operation include four Silver Stars, 26 Bronze Stars, five Purple Hearts and 71 Army

Commendation Medals. The courage and tenacity on Objectives Lynx and Cobalt by an outnumbered and outgunned force set the example for follow on Ranger operations in the decade long war.[37]

A DECADE AND MORE OF RANGERS IN COMBAT

The parachute assault onto Rhino at the beginning of OEF was the first large Ranger operation since the Battle of Mogadishu. The Rangers met with success during the invasion into Afghanistan and, along with the other US Special Operations forces, played an integral part in eventually overthrowing the Taliban government. They also participated in the deadliest fight of Operation Anaconda in March 2002 at Takur Ghar in an effort to rid the Shahi-Kot Valley of Al-Qaeda.

The operations conducted by the Rangers in both OEF and OIF are numerous and too large of a scope to describe for this work. However, a snapshot of their missions as shown in this chapter defines the tenacity and courage displayed for over a decade of combat. By the time 2001 came about this Special Operations Force forged with the spirit of discipline and professionalism as envisioned by their founding fathers had come to fruition. They were a force to be reckoned with.

Before OEF and OIF the Rangers, and the US Army as a whole for that matter, had been involved in short duration (albeit intense at times) operations in short periods of time. OEF and OIF would prove their ability to sustain in combat in short durations for great periods of time. As they progressed through this war their feats and accomplishments were often recognized with amazement and awe. For instance, it's a little known fact, that the Rangers would conduct three combat parachute operations in Afghanistan and two in Iraq since the GWOT began. History has yet to catch up with this incredible set of accomplishments and the chapter has yet to be closed.

During a decade of action, the Rangers have had less than half the casualty rate suffered by conventional forces as the result of an elite training regimen and quintessential leadership principles and manning. In support of this statistic, the Rangers developed more aggressive and effective techniques for treating combat casualties, which substantially reduced the death rate. As discussed earlier the rest of the army has adopted many of these techniques.

Sean Naylor understood the benefits of Abrams Charter and the Ranger Creed provided by the modern day Ranger when he wrote his "Not a Good Day to Die- The Untold Story of Operation Anaconda." In this article he writes, "… But it was the Ranger Regiment that provided the largest set of shared experience that connected the leaders gathering at Bagram [Afghanistan]. The Regiment falls under US Special Operations Command but is really an elite airborne infantry force that links the light infantry and special operations communities. Unlike Delta or Special Forces, into which troops tend to disappear for the rest of their careers, soldiers often rotate between the Ranger Regiment and the Army light infantry divisions. So it was that many Mountain [10th Mountain Division] and Rakkasan [3rd Brigade of the 101st Airborne Division] officers and senior NCOs had served together in the Rangers. This was a massive slice of good fortune. The 75th Ranger Regiment is a tight community of warriors whose ethos is summed up in the Ranger Creed. There are 241 words in the Ranger Creed, and every Ranger is required to learn them all by heart. But the Creed's essence is encapsulated in six of them: 'Never shall I fail my comrades.'"[38]

As the years unfolded in both OIF and OEF the Rangers usually deployed for three months at a time, and served as a special capability for important operations. The 75th Rangers, who belong to SOCOM, and work for JSOC (Joint Special Operations Command) often act as backup for the elite American commandos (Delta Force and Seal Team 6) and also carry out many operations on their own. During a typical three month tour, the Rangers might average three to four missions in the form of patrols, raids, air assaults, mounted infiltrations behind enemy lines, complex urban assaults and rescue operations a day, each one resulting in three to four enemy killed and seven to eight captured, along with large quantities of weapons and documents.

While at war the Regiment also continued to train, recruit, assess the next generations of Rangers. To date more than 80% of the active duty Rangers have conducted multiple combat deployments and many are on their sixth or seventh rotation since the beginning of 9/11.

Rangers continue to conduct the full array of infantry and Ranger-type missions including:

- Airfield Seizures using both special operations and light infantry methods to -
  * Establish air-landing capabilities for follow-on forces
  * Establish trans-load sites for precious cargo (PC)
- Special Operations Raids to:
  * Destroy high payoff targets
  * Recover or rescue precious cargo (PC)
- Military Operations in Urban Terrain
- Employing precision marksmanship against high value targets (HVT)
- Using advanced combatives like hand-to-hand techniques[39]

Throughout the war on terror the Army maintained the Regiment at a high level of readiness. Each battalion remains deployable anywhere in the world within 18 hours. To maintain this readiness, the Rangers constantly train in a wide array of areas encompassing techniques required for all environments.[40] Most of this training is conducted at night with a focus on quick reaction and crisis action planning.

Often the Rangers are brought in to help JSOC with searches for Islamic terrorists. In these cases the Rangers spend most of their time patrolling, or on stakeout, noting everything and developing a web of information that will catch the bad guys. Of note was the raids on Saddam's sons Odai and Qusai when on July 22, 2003 the 101st Airborne Division's 2nd Brigade under the command of then-Col. Joe Anderson (a serving Lt. Gen. and Ranger combat veteran) directed the killing of the oppressive brothers. Additionally, the killing of Abu Musab al-Zarqawi, an Al-Qaeda leader in Iraq had a series of Ranger missions that helped locate this terrorist leader. Zarqawi died in a targeted killing on June 7, 2006, while attending a meeting in an isolated safe house approximately eight km north of Baqubah. Two United States Air Force F-16C jets dropped two 500-pound guided bombs on the terrorist safe house. The Rangers helped identify this important high payoff target.

While the combat parachute assault on Objective Rhino on October 19, 2001 and the actions involving a small Ranger Quick Reaction Force (QRF) at Takur Ghar on March 3-4, 2002 were the early highlights of the war in Afghanistan, these were not the only missions conducted by the Regiment. Long range combat patrols in Desert Mobility Vehicles (DMV), QRF raids to destroy a target during a single cycle of darkness and other mission sets tailored for short duration but extremely lethal would become the hallmark of the Rangers in both theaters.

With great success comes great rewards. Rangers have proven themselves very courageous during the two wars in the Middle East. As of this writing there has been a combined total of sixteen Medal of Honor recipients in the two wars combined- (twelve in Afghanistan and four in Iraq. The GWOT has also seen the first time since Vietnam that a Medal of Honor has been awarded to a living recipient. The first was awarded to Staff Sergeant Salvatore Giunta of the 2nd Battalion, 503rd Infantry (Airborne) and the second to a member of 2nd Battalion, 75th Rangers; Leroy Petry. Petry was shot through both legs when lost his right hand while throwing a live grenade away from his fellow soldiers.

Even though the Rangers are continuously engaged in what became termed "Task Force" operations in OIF and now continuously in OEF, they continued to evolve as an organization both in structure and in operational Tactics, Techniques and Procedures (TTP). For instance, due to the changing nature of warfare and the need for an agile and sustainable Ranger Force, the Regimental Special Troops Battalion (RSTB) was activated on July 17, 2006. The RSTB is responsible for many of the special Ranger support missions to include sustainment, intelligence, reconnaissance and maintenance missions which were previously accomplished by small detachments assigned to the Regimental headquarters and then attached within each of the three Ranger battalions.

As 2010 came to an end, the War in Iraq followed suit. Arguably all of the large terrorist cells had been neutralized by mid-2010 and the Iraqi forces were in the security lead throughout the country. The terrorist leaders had been killed or captured and the Iraqi population supported the Americans and opposed the terrorist. This was a big change since the successful surge in 2007 when Gen. Ray Odierno became the US Forces-Iraq Commander. Unfortunately, with the abrupt departure of US forces the Sunni-Shite battle filled the gap of the terrorist cells and challenges in that war torn country continue today.

Much of the short term successes achieved in terminating the war, or at least changing the face of battle, can be attributed to US Army Rangers, both in the Regiment and as part of the forces that filled the ranks of the conventional Army. With great success came great sacrifice but that sacrifice wasn't left unrecognized. The lines of how the conventional force and the special operations forces operated were narrowed making Army that much better.

Abrams Charter had proven itself in the GWOT. The vision Abrams and his initial disciples, Leuer, Rokosz, Nightingale, Currie and others was spot on and thirty-plus eight years later the nation is reaping the benefits. No Army in history has been more professional or more formidable. The US Army Rangers forged as a Special Operations Force clearly had a large impact on this success.

Future role of the Rangers in a post Afghanistan/Iraq world.

Those who study warfare only in the light of history think of the next war in terms of the last. But those who neglect history deprive themselves of a yardstick by which theory can be measured.

Cyril Falls, author *The Nature of Modern War*

The future of any military force, regardless of its type or purpose, is dependent on the future of warfare. What will be wars like in the coming years? Will they emulate the counterinsurgency, low intensity short duration conflicts of the wars just witnessed in Afghanistan and Iraq? Or will they be all-out tank battles like those envisioned with the Chinese, the North Koreans or some other large highly resourced adversary? Or will they be of a different flavor all together like a net centric war where controlling information and wealth are the goal? No one really knows but there is much speculation. Conventional wisdom would say that the short answer should be 'yes' to all questions above – at least from a military standpoint. Recognizing that all options and scenarios are possible makes a nation ready both militarily and culturally for any potentiality or series of potentialities.

An Army must train to fight the next war based on the lessons of the last war and a projective analysis of the future fight. Many theorems have been developed in the study of the wars of the future. Often we hear how nations should never ever fight the last battle. But the future is indeed dependent on the past in many ways. Much angst with fighting the last fight stems from our culture's emphasis not to fight another "Vietnam." While Vietnam and its aftermath was a failure in many ways, OIF and OEF showed promising signs of success in the resolution of conflict and war termination transition of responsibility. This was indeed the case until both were abruptly abandoned by the US and the coalition partners to let the indigenous security forces vie for themselves. The future of warfare and strategic success depends on the emulation of those process and procedures that proved successful and to shun those that were not.

What is often ignored through fear of rehashing and mirror imaging the past are the lessons learned through experience. It's a comfort zone to replicate the next fight on what has occurred recently. This is true, especially if the most recent war is still ongoing like in the war in Afghanistan and the aftermath of Iraq. Spending billions of dollars in an effort to develop a formidable force at a huge sacrifice makes a nation default to what it has and what it knows when it comes to military tactics, techniques and procedures.

The massive amount of life lost coupled with the hardships of personal lives changed forever by separation, wounds and post-traumatic stress syndrome (PSTD) in the most

Today's Rangers Jumping from a C17 Starlifter in 2009.

Rangers exiting a Little Bird.

recent wars makes one accept that what is in the inventory militarily already is good enough. It is human nature to default to what one knows; especially if it works and our Army today works. This attitude is a great benefit while at the same time a great risk.

We already discussed that the gap between what it's like to be a Ranger and a conventional soldier is closing. Because of the desire not to fight another Vietnam, the Abrams' Rangers of the 1974 era saw a need for an elite force to fight the high end spectrum of combat. This elite force would also have the rest of the Army emulate it in the form of standards and professionalism and there is a significant "return on investment" regarding the spread of the Ranger ethos and warrior spirit throughout the Army as a whole.[1]

As we look to the future, the challenge is for the Rangers of today is to regain their footage in maintaining the progressive and expansive edge to keep the tactical ideas fresh and the rest of the Army curious as to how to keep up. Stagnation for any Army at any point in history where a nation has proven its wealth militarily is risky. While recent successes are mixed, our enemies continue to find ways to overcome the already existing capabilities in our force. Change is necessary if for no other reason than to survive and maintain the competitive edge.

This chapter takes a general look at follow-on specialized and global training for our current Ranger force by reviewing current force structure and possible roles for the future. Reviewing the history of the Rangers in this work has given the reader the chronological understanding of their evolution. By examining the implementation of the Creed and the Charter has taken this understanding to the next level showing the reader the impact this force had on the Army today. A peek into what may be in store for Ranger Force capabilities in the future is the natural way to bring this book to its end.

THE CHANGING FORCE

There is little doubt that Abrams' Rangers changed the Army. As the Army adapted and changed by learning lessons from the wars in Iraq and Afghanistan, the Regiment of Rangers continued to evolve. With the requirements posed on the operational force, the 75th Ranger Regiments Headquarters and its three Ranger Battalions have adjusted in the past decade to the needs presented by persistent war.

The last chapter provided an explanation of the addition of the Ranger Special Troops Battalion (RSTB) to provide special skill sets. Logistical support has always been a challenge for the austere makeup and employment of the Ranger force. Expanding these logistical support functions with the formulation of the RSTB has provided the Regiment increased operational capabilities in its effort to sustain combat operations.

The RSTB and its four organic companies is chartered to conduct command, control, communications, computers, intelligence, surveillance and reconnaissance functions in support of an employed Task Force combat operation.[2] These skill sets are necessary to enable the execution of Joint Special Operations anywhere in the world. Additionally, the RSTB provides qualified, trained and ready Rangers in order to sustain the Ranger Force. This function is unlike any STB in the Army.

One of the most beneficial changes that benefited each Ranger was the inclusion of high level medical training. To increase the ability to provide battlefield medical care Ranger medics are required to attend the pre-special operations combat medical course (PSOCM). This course trains and qualifies medics to manage trauma, prepares combat wounded for evacuation and provide basic medical care at the battlefield location of the wound. It provides training for a full array of basic life support to include pharmaceutical calculations; anatomy; physiology; pathophysiology; medical terminology; exam techniques; documentation; airway management; assessment; operating room minor surgical skills and procedures and a number of other skills making the Ranger medic increasingly capable in saving lives on the battlefield.

In 2007 the Reconnaissance Detachment (RRD) transformed into the Ranger Reconnaissance Company (RRC). While the Rangers remained at war in the early 2000s they saw the need for a larger more capable reconnaissance force due to its extensive training and unique capabilities to conduct special reconnaissance and close target reconnaissance (CTR) operations. Providing worldwide reconnaissance preparation of the battlefield allows the Rangers to remain relevant on a global scale.

Expanding the global ability of the Regiment requires a full array of communication capabilities. The Ranger Communications Company (RCC) has the equipment and skill sets for the Regiment to provide communication and command and control support. The need for interoperability in communication requirements across the joint environment while meeting the additional communication requirements of other special operations task forces made the addition of the RCC absolutely necessary.

The RSTB's Military Intelligence Company (MICO) offers the force an ability to conduct the full array of intelligence functions including SIGINT, HUMINT, IMINT, and all source analysis operations in support of combat operations. The expanded capabilities of the MICO provides the Ranger force an ability to synchronize intelligence, gather and analyze full motion video, and provide advanced collection and analysis capabilities. The addition of organic analysts and operators to an already robust military intelligence command was based on requirements identified in the small unit raids and strategic to tactical high value target "hits" performed in the past decade in both theaters.

Finally, a key part of the RSTB and unlike other STBs in the Army, is the Ranger Selection & Training Company (RSTC). The purpose of this company is to provide an ability to assess and evaluate Ranger candidate so that the force is properly manned. The programs of instructions (POI) includes the Ranger Assessment and Selection Program (RASP 1 and 2), and Small Unit Ranger Tactics (SURT). RASP 1 assesses, trains, and identifies soldiers of the rank of E-5 and below. RASP 2 conducts the assessment and selection for soldiers E-6 and above. These courses are mandatory for those aspiring to wear the Ranger scroll. What was once called Pre-Ranger is now called Small Unit Ranger Tactics (SURT). SURT prepares members of the Regiment who are not yet Ranger qualified for successful completion of the United States Army Ranger School.[3]

One command not formally part of the Regiment but an absolute necessity to its success is the Ranger Training Brigade (RTB). The RTB has evolved continuously over the last 60 plus years since the first four companies graduated from the six week course on November 13, 1950 by the then-Ranger Training Command. These newly Ranger qualified soldiers were expecting to be awarded an insignia representing the white and read on black scroll worn by the World War II Ranger Battalions. Instead they received gold on black arced insignia with the word RANGER in the arc; the very first "Coveted Black and Gold." The birth of the Ranger Tab may have not been that exciting to the newly tabbed Rangers some 60-plus years ago, but today the Black and Gold Tab is indeed coveted in today's Army.

Today the Ranger course is 61 days in length averaging 19.6 hours of training each day for seven days a week.[4] The three phases include the "city" phase which is held in Fort Benning, Georgia, the Mountain phase which takes place in Dahlonega, Georgia, and the "swamp" phase held in Eglin Air Force Base, Florida. Over the years there was a "desert" phase as well which was initiated by Ken Leuer and was held in Dugway Proving Grounds, Utah and periodically moved to Fort Bliss, Texas. For whatever reason that phase fell by the wayside and the three phases that exist today produce some of the best Rangers our Army has ever seen. As the Army changes the need for skills our Rangers should have to change as well. However, the basic patrolling skill and movement techniques found in the Ranger school curricula are the staples for all small unit leaders.

ANALYZING WHAT COULD COME NEXT

The Rangers have to be ready for the next fight for there is no doubt that when the guns start firing the nation will beckon them to the battlefield. Being prepared takes some foresight and the history of the Ranger command is replete with visionaries and resource minded leaders. In 1974 KC Leuer had a vision of a performance oriented trained and disciplined force; a force that was an elite fighting force that the rest of the Army would emulate. That vision, while simple in today's terms, was profound and changed the course of our Army

As Leuer moved on, other Battalion and Regimental Commanders made their mark on the Rangers and the Army. In the late 1990s, Stan McChrystal and Ken Keen were examples of Ranger visionaries.[5] It is clear now that what McChrystal and Keen implemented as Ranger leaders and followed through by Joe Votel completely prepared the Regiment for the Global War on Terror (GWOT).

The focus on counter-terrorism training and operational planning was the genesis for how the Rangers evolved. Information attained from web-based situational awareness venues leading to a common operating picture (COP) allowed the Rangers to better see themselves and see the enemy. A balanced integration into both the special operations community and keeping a wedge in the Big Army's Infantry Center keeps the Regiment relevant in the global landscape. Modernizing weaponry and the implementing the arms room concept ensures the Rangers are lethal and keep their combative edge. Professionalizing the medical and Ranger First Responder training helps keep Rangers alive during the "Golden Hour" after a wound occurs. These and other wide reaching changes were courageously made by Ranger leaders to prepare the Regiment for the coming fights in both OEF and OIF. What is remarkable is that McChrystal, Keen and Votel had the vision to make these directed changes without obviously knowing what was on the horizon in the next five to ten horizon.

The 1993 Mogadishu fight, the 1994 Haiti planned invasion and various Bosnian and Kosovo deployments in the late 1990s were basically all the Ranger leaders had to go on to make visionary plans for the future Ranger requirements at that time. Somehow the quantum leap to a major counter-terrorist, small unit raid focused, and rapid deployment capability forced was envisioned by these leaders and the results were astounding. There is no doubt sweeping change and implementation is difficult. It takes moral courage to make significant changes to any organization especially where the risks are high and the payoff is uncertain.

As the force looks to the future, and to keep pace on the cutting edge of tactics, operations and strategic thought, the Rangers have to maintain the extraordinary hiring process developed over the years. Bringing in experienced, intelligent and balanced warriors as Regimental Commanders and already qualified battalion commanders from the conventional Army as Ranger Battalion commanders will provide that potentiality for visionary thought.

To challenge ourselves in visionary thought we must ask ourselves how we might try to think differently about the future for military planning purposes. A useful way to begin this process is to identify trends—ongoing processes that have considerable momentum—that are likely to continue into the future with relatively limited, or only gradual, changes. Demographics is one of them.[6]

For instance, the demographic decline and collapse of public health in various countries like Russia has a global impact. The Soviet Union unemployment rate declined when the wall was dismantled two decades ago. Its society has had a hard time rebounding since. Life expectancy issues coupled with economic woes makes a resurgence of Russian national power in the next 20 years unlikely. This potential unrest, however, could have an impact on rogue terrorist involvement and great instability in that region. Military leaders need to explore how these factors, for instance, have an impact on future requirements for a force such as the Rangers.

The aging population contraction in Europe and Japan is also insightful. This demographic factor makes Europe and the Far East unlikely centers of power in the future. The position of Europe is particularly of interest, since the countries across the Mediterranean in the Levant are growing in population. An already large Islamic population in Europe may have an impact on the potential for fundamentalism or even civil unrest due to racism and religious misunderstandings. Fighting Islamic fundamentalists in an Arab centric nation state is different from attempting to neutralize this kind of terrorist effort in Western societies. Forces like the Rangers and other special mission units need to develop tactics, techniques and procedures for operating along the full spectrum of conflict in areas where their involvement would be perceived as sensitive at best.[7]

As nations of power, and those that have habitually seen as having power, vie for global positioning the stability of the already existing nuclear arsenal becomes at risk. While the potential for nuclear terrorism is uncertain, the Rangers must master the skills to defeat a terrorist element known to have acquired weapons of mass destruction. How this fits into the future force and its training regimen is yet to be seen.

China is the wild card in all of this. It is clearly an economic giant with a great population base and aspirations for more. Its status as a world's superpower in competition for resources with the US could have an impact on world stability and no doubt require attention militarily in areas well beyond the high intensity tank battles envisioned in a US-Chinese war. The Regiment needs to keep a keen eye on this already existing adversary.

Driven by power, patriotism, and curiosity can turn even the most intelligent and informed men to new scientific discoveries into weapons of greater capability and those of mass destruction. A predisposition to defend oneself and a desired way of life inherently justifies research and development of new weapons, however horrendous their effects.[8] Situational awareness of these technological evolutions and posing tactics against their proliferation and use is what a force like the Rangers needs to develop well in advance of any prospect of it happening.

Population growth and its impact on security causes concern for global unrest and the impact on a dwindling pool of resources to support an explosion in population growth. Global warming, poverty and the emergence of diseases are daunting challenges as well.

Colonel Ralph Pucket is the Honorary Colonel of the 75th Ranger Regiment today. He is intimately involved in all the Rangers do to include taking care of the warrior families. (Source: Caraccilo)

Why Rangers Really Fight. (Source: Currie)

Anyone can brainstorm ideas for the makeup of the next Ranger Force and its capabilities for the future.[9] Unconstrained course of action development focusing on organizational structure, manning and resource requirements to fulfill the "shoot, move, and communicate" requirements can result in everything from leviathan boots to exoskeleton robotic body armor light enough to offer 100% protection without reducing the agility of the "soldier as a system." But are they realistic and can they be achieved, and are these unconstrained requirements in the extreme cases even warranted? It is up to the leaders today to look out smartly on all that is happening and could happen to define the way ahead for the Rangers.

FAMILY READINESS

Before ending this work one of the ancillary effects and perhaps one of the most important but little known contribution of the first Ranger Battalions to the Army was the concept of the Family Readiness Group (FRG). The Family Readiness Group is a unit's organization of family members in support of that organization's mission. Its purpose is to provide a family support structure for an organization. This type of support is especially important while a unit is deployed.

Today the Family Readiness Group is very much an institutionalized part of the Army. In fact, unit commanders at every level are required by Army regulation to facilitate the establishment and maintenance of such an organization. Such organizations did not exist in the Army in 1974 when Abrams' Rangers were formed.[10] Although families of soldiers have been helping each other since the days of the frontier, no formal structure existed. The nature of the formation and the basing of Abrams's Rangers created the environment for such an organization.

The first Ranger battalions were established as separate units on posts dominated by much larger divisional units. The 1st Ranger Battalion was activated at Fort Stewart, Georgia, which was the home of the 24th Infantry Division. Fort Lewis, Washington, was chosen as the home of the 2nd Battalion. Fort Lewis was also the home to the 9th Infantry Division. Since their inception, the Ranger Battalions were envisioned as strategic assets, they were not under the command of the infantry divisions at their respective installations. Consequently, the Ranger battalions found themselves as outsiders. This drove the need for the families to bond in mutual support.[11]

Lieutenant General (Retired) Lawson W. Magruder III (former commander of B Company, 2nd Ranger Battalion) expounded on this topic in an interview 2003:

> Then, as now, the 82nd Airborne Division was considered one of the Army's best divisions. Magruder said that the 82nd Airborne Division did not have any such program as an FRG, but one was in place when he arrived to command B Company, 2nd Ranger Battalion.
>
> One of the wives assigned to the Ranger Battalion at the time wrote about her experience with what became known as Family Readiness Groups. She wrote: This was my first assignment as a "wife/dependent" and watching how they [the battalion commander and his wife] maintained that big family has always been a source of inspiration for me – I still tell new wives to the military how great it can work.
>
> Everyone was important and everyone mattered. As wives we looked out for each other constantly, all the way from…Cdr's wife down to the newest and youngest wife. This wife went on to write that she continues to pass on the lessons she learned

Plankholders Todd M. Currie and KC Leuer in 2014. (Source: Currie)

Modern Day Special Operations Soldier conducting a parachute jump in Africa.

as a spouse of a Ranger. She described practices of those early wives groups that are found in today's formal Army Family Readiness program. Empirical evidence such as these examples certainly suggests that Family Readiness Groups are one of the practices of the Ranger Battalions that migrated to the rest of the Army."[12]

Like the performance oriented training and standardized training processes introduced by KC Leuer and the rest of the 1st Battalion Plankholders in the spring of 1974, the standard for taking care of the very important population each Ranger strives to defend, the American Family and its way of life, had found its way into an organization born of Ranger values and provided to the rest of the Army.

Epilogue

They've got us surrounded again, the poor bastards.

General Creighton W. Abrams

You just experienced US Army Rangers A to Z. While negotiating this work, a work that resulted in the forging of an elite special operations unit, the total impact on the establishment and development of the modern day Ranger, the Ranger and their families unit may not be fully realized.

Note that in the trail end of the book, the founding fathers like Leuer, Rokosz, Currie and Nightingale are mentioned less as they moved off to senior jobs and eventually retired. However, as the Rangers, that they founded, grew into a formidable force and the best light infantry force the world has ever known, their legacies have lived on.

The successes the Rangers accomplished are numerous and this work attempted to capture the salient moments in history that define historical and modern day Rangers. Focusing on the development of the post-Vietnam era Ranger and the psyche of the Army in the mid-1970s the reader can understand how the Rangers evolve into what they are today.

Today's US Army Rangers. (Caraccilo)

Todd Currie and KC Leuer in 2009 at the 75th Ranger Rendezvous in Ft. Benning, GA. (Source: Todd Currie)

Plankholder Colonel (Retired) Keith Nightingale being award the French Legion of Honor at St Mere Eglise, Holland in 2013. (Source: Nightingale)

Some critics have maintained that the Rangers are unnecessary because there is nothing the Rangers do that cannot be done by the conventional infantry. While there is some validity to that statement, the set of circumstances for developing the Ranger and the Ranger unit does not exist in the same vain in a conventional force. Manning and equipping a small, focused force like the 75th Ranger Regiment comes with a great deal of benefit. Because it draws from the Big Army and has the dedicated resources to train as it does, it can be manned, resourced and equipped with the best of the best. It is designed that way so there is no comparison and the lack of comparison has nothing to do with the abilities of the conventional infantry. In fact, it's the Rangers who set the bar for the rest of the infantry to strive to emulate.

As of this writing the Army has recognized how important the US Army Ranger and the 75th Ranger Regiment is to the security of our nation as they continue to deploy them on missions around the world. The Army and the men and women who fill its ranks continue to be a snapshot of our society and will continue to represent the American culture and the American way of life. The future of our Ranger ranks will be a cross reference representation of race, ethnicity and even gender.

The Charter and the Creed are lasting because of how they set the stage for the modern day force. The lessons learned are numerous and beyond the scope of this work. However by anyone's standards, the Ranger battalions of the 75th Ranger Regiment are the most elite and proficient light infantry in the world. As long as they continue to remain focused on their mission and to keep in mind the lessons learned of the past, the Regiment will remain capable of fulfilling the Army Chief of Staff's initial Charter to the Regiment, that of Gen. Abrams, of fighting anytime, anywhere, against any enemy, and winning.

Appendix A

Ranger Hall of Fame

The man who conquers in war is the man who is least afraid of death.
Alexi Kuropatkin, Russian General and Minister of War, 1848-1925

RANGER HALL OF FAME (RHOF)[1]

The Ranger Hall of Fame (RHOF) was formed to honor and preserve the spirit and contributions of America's most extraordinary Rangers. The members of the Ranger Hall of Fame Selection Board take particular care to ensure that only the most extraordinary Rangers are inducted; a difficult mission given the high caliber of all nominees. Their precepts are impartiality, fairness, and scrutiny. Inductees were selected impartially from Ranger units and associations representing each era or Ranger history. Each nominee was subjected to the scrutiny of the Selection Board to ensure the most extraordinary contributions are acknowledged.

The selection criterion is as unique as the Ranger history. To be eligible for selection to the Hall of Fame, a person must be deceased or have been separated, or retired from active military service for at least three years at the time of nomination. He must have served in a Ranger unit in combat or be a successful graduate of the U.S. Army Ranger School. A Ranger unit is defined as those Army units recognized in Ranger lineage or history. Achievement or service may be considered for individuals in a position in state or national government after the Ranger has departed the Armed Forces.

Honorary induction may be conferred on individuals who have made extraordinary contributions to Ranger units, the Ranger foundation, or the Ranger community in general, but who do not meet the normal criteria of combat service with a Ranger unit or graduation from the U.S. Army Ranger School.

Each inductee is presented with an engraved, specially cast bronze Ranger Hall of Fame medallion, suspended from a red, white and blue ribbon. The medal signifies selfless service, excellence and remarkable accomplishment in the defense of the nation and to the highest ideals of service.

Inductee	Title	Year	
Abood, Edmond P.	COL	7th Annual (1999)	
Abrams, Creighton W.	GEN	Inaugural Year	1992 Inaugural Year
Acebes, William H.	CSM	11th Annual (2003)	
Acker, John A.	TSGT	12th Annual (2004)	
Alderman, Joe C.	1SG	4th Annual (1996)	
Alison, John	MG	14th Annual (2006)	Honorary induction
Allen, Warren E.	MAJ	5th Annual (1997)	
Altieri, James J.	MAJ	17th Annual (2009)	
Anderson, William E.	PFC	5th Annual (1997)	
Anton, William T.	LTC	17th Annual (2009)	
Antonia, Keith P.	LTC	17th Annual (2009)	
Arnbal, Anders K.	1SG	3rd Annual (1995)	
Atkins, Albert	SFC	9th Annual (2001)	
Auger, Sr., Ulysses G.	T4	13th Annual (2005)	
Baker, AJ "Bo"	COL	17th Annual (2009)	
Barber, Harold L.	MAJ	14th Annual (2006)	
Bargewell, Eldon A.	MG	19th Annual (2011)	
Barton, James	SGM	20th Annual (2012)	
Basil, Albert E.	Father	4th Annual (1996)	
Bates, Jared L.	LTG	18th Annual (2010)	
Batts, Aubrey M.	MSG	3rd Annual (1995)	
Bayne, Robert D.	LTC	12th Annual (2004)	
Beach II, Charles E.	COL	3rd Annual (1995)	
Beach, George I.	MSG	15th Annual (2007)	
Beckwith, Charles A.	COL	9th Annual (2001)	
Bennett, Harold G.	SSG	12th Annual (2004)	
Berg, Gilbert M.	MSG	15th Annual (2007)	Honorary induction
Biggs, Tom	SGT	15th Annual (2007)	Honorary induction
Birch, Alfred G.	CSM	20th Annual (2012)	
Black, Robert W.	COL	3rd Annual (1995)	
Blair, Melvin	LTC	11th Annual (2003)	
Block, Walter E.	CPT	6th Annual (1998)	
Block, William D.	1SG	18th Annual (2010)	
Boatman, Roy	SFC	15th Annual (2007)	
Bowman, Donald C.	LTC	19th Annual (2011)	
Branscomb, Clarence	SSG	4th Annual (1996)	
Brasher, Al	SMJ	18th Annual (2010)	
Brown, Roger B.	CPT	12th Annual (2004)	
Brownlee, Les	HON	11th Annual (2003)	
Bryan, Robert L.	SGT	14th Annual (2006)	

Inductee	Title	Year	
Bucha, Paul W.	CPT	7th Annual (1999)	Medal of Honor
Buckelew, Alvin H.	LTC	15th Annual (2007)	
Bukaty, Alford M.	SFC	17th Annual (2009)	
Burkett, B.G.	Mr.	13th Annual (2005)	Honorary induction
Burnell, Richard C.	CSM	4th Annual (1996)	
Butler, William	SFC	8th Annual (2000)	
Callaway, Howard H.	HON	11th Annual (2003)	Honorary induction
Caro, Henry	CSM	1st Annual (1993)	
Carpenter, Gary R.	CSM	6th Annual (1998)	
Carr, Robert L.	SGT	10th Annual (2002)	
Carrier, Charles L.	MAJ	14th Annual (2006)	
Caruth, Charles W.	CW4	12th Annual (2004)	
Castonguay, Romeo J.	MSG	11th Annual (2003)	
Cavazos, Richard E.	GEN	Inaugural Year (1992)	
Cavezza, Carmen J.	LTG	18th Annual (2010)	
Chaney, George	CSM	4th Annual (1996)	
Channon, Robert I.	COL	4th Annual (1996)	
Church, Benjamin	CPT	Inaugural Year (1992)	
Cicuzza, Sisto	SGM	9th Annual (2001)	
Clark, Jr., William C.	SGT	3rd Annual (1995)	
Clement, Albert	MAJ	8th Annual (2000)	
Clemons, Jr., Joseph G.	COL	7th Annual (1999)	
Cobb, Autrail	CSM	10th Annual (2002)	
Collier, James H.	CSM	10th Annual (2002)	
Connell, Kevin P.	CSM	14th Annual (2006)	
Conrad, George D.	CSM	16th Annual (2008)	
Corley, John T.	BG	11th Annual (2003)	
Cournoyer, Joseph R. M.	SGM	7th Annual (1999)	
Crabtree, Ormand B.	SGT	4th Annual (1996)	
Dammer, Herman W.	COL	5th Annual (1997)	
Daniel, Jr., John S.	LTC	7th Annual (1999)	
Darby, William O.	BG	Inaugural Year (1992)	
Davis, Fred E.	SGM	2nd Annual (1994)	
Dawson, Francis	COL	Inaugural Year (1992)	
Dean, Edwin L.	LT	17th Annual (2009)	
Delorey, Donald W.	CPT	5th Annual (1997)	
Devlin, Gerald M.	MAJ	3rd Annual (1995)	
Doane, Stephen H.	1LT	5th Annual (1997)	Medal of Honor
Dodge, Russell E.	1SG	17th Annual (2009)	
Dolan, Gary E.	LTC	19th Annual (2011)	

Inductee	Title	Year	
Dolby, David C.	SSG	1st Annual (1993)	Medal of Honor
Donovan, William "Doc"	CW4	7th Annual (1999)	
Dorman, Elwood L.	PFC	16th Annual (2008)	
Downing, Wayne A.	GEN	7th Annual (1999)	
Dubik, James M.	LTG	20th Annual (2012)	
Dushane, Cyrille J.	SSG	5th Annual (1997)	
Dye, Raymond N.	PFC	18th Annual (2010)	
Dyke, Charles W.	LTG	15th Annual (2007)	
Eaton, Richard J.	BG	6th Annual (1998)	
Edlin, Robert T.	LT	3rd Annual (1995)	
Edmunds, John R.	CSM	18th Annual (2010)	
Ehrler Jr., Richard S.	SGT	6th Annual (1998)	
England, Steven R.	CSM	19th Annual (2011)	
English, Jr., Glenn H.	SSG	10th Annual (2002)	
Evans, Warren	CPT	4th Annual (1996)	
Fike, Jr., Emmett E.	CPL	8th Annual (2000)	
Fletcher, Larry A.	CSM	4th Annual (1996)	
Foss, John W.	GEN	3rd Annual (1995)	
Franklin, John W.	SGT	15th Annual (2007)	
Frost, Hubert H.	SGM	4th Annual (1996)	
Gagnon, Joseph R.	CSM	18th Annual (2010)	
Galvin, John R.	GEN	16th Annual (2008)	
Gardner, Gregory E.	CPT	2nd Annual (1994)	
Gardner, James	1LT	14th Annual (2006)	Medal of Honor
Gary, Robert P.	LTC	17th Annual (2009)	
Gates, Julius W.	SMA	8th Annual (2000)	
Geer, Robert J.	SSG	9th Annual (2001)	
Gentry, Neal R.	CSM	1st Annual (1993)	
Gergen, Theron V.	CSM	11th Annual (2003)	
Getz, Charles E.	BG	10th Annual (2002)	
Gividen, George M.	CPT	12th Annual (2004)	
Gooch, Christopher M.	LTC	12th Annual (2004)	
Gordon, Gary I.	MSG	3rd Annual (1995)	Medal of Honor
Gore, William E.	LTC	2nd Annual (1994)	
Gosho, Henry	SSG	5th Annual (1997)	
Grange, David L.	BG	13th Annual (2005)	
Grange, Jr., David E.	LTG	Inaugural Year (1992)	
Grissom, William C.	LTC	2nd Annual (1994)	
Hale, Nathan	CPT	1st Annual (1993)	
Hall, Glenn M.	CPL	Inaugural Year (1992)	

Inductee	Title	Year	
Hanks, Tom	Mr.	14th Annual (2006)	Honorary induction
Harris, Randall	1SG	3rd Annual (1995)	
Heath, Mayo S.	1LT	2nd Annual (1994)	
Heckard, Joe M.	CSM	20th Annual (2012)	
Herbert, James A.	BG	12th Annual (2004)	
Herring, Thomas E.	SGT	14th Annual (2006)	
Hirabayashi, Grant J.	TSGT	12th Annual (2004)	
Holland, James	COL	16th Annual (2008)	
Hopkins, James E.	CPT	10th Annual (2002)	
Horvath, George	CSM	12th Annual (2004)	
Howard, Robert L.	COL	13th Annual (2005)	Medal of Honor
Hunter, Charles N.	COL	1st Annual (1993)	
Jacks, Danny L.	SSG	13th Annual (2005)	
Jackson, James T.	MG	18th Annual (2010)	
Jakovenko, Vladimir	SGM	11th Annual (2003)	
Janis, Norman	PFC	4th Annual (1996)	
Johnson, Caifson	LTC	7th Annual (1999)	
Karbowski, Stanley	1SG	15th Annual (2007)	
Katz, Warner	SSG	8th Annual (2000)	
Kelly, Sean T.	1SG	19th Annual (2011)	
Kelso, Michael A.	CSM	18th Annual (2010)	
Keneally, John T.	COL	13th Annual (2005)	
Kernan, William F.	GEN	14th Annual (2006)	
Kerrey, J. Robert	SEN	3rd Annual (1995)	Medal of Honor, Navy
Kimsey, James V.	MAJ	13th Annual (2005)	
Kingston, Robert C.	GEN	4th Annual (1996)	
Kirshfield, William	SGT	8th Annual (2000)	
Klein, Gerald E.	CSM	20th Annual (2012)	
Kness, Lester E.	CPT	10th Annual (2002)	
Knight, Jack L.	1LT	13th Annual (2005)	Medal of Honor
Kuhn, Jack E.	SSG	2nd Annual (1994)	
Labrozzi, Anthony	COL	11th Annual (2003)	
Lacy, Joseph R.	Monsignor	4th Annual (1996)	
Langmack, Steven	SFC	20th Annual (2012)	
Law, Robert D.	SP4	Inaugural Year (1992)	Medal of Honor
Laws, Jr., Charles R.	1SG	10th Annual (2002)	
Lawton, John P.	COL	5th Annual (1997)	
Laye, Jesse G.	CSM	13th Annual (2005)	
Leandri, Richard A.	Mr.	Inaugural Year (1992)	Honorary induction
Lehew, Donald L.	SFC	7th Annual (1999)	

Inductee	Title	Year	
Lemon, Peter C.	SGT	2nd Annual (1994)	Medal of Honor
Lemonds, Gary L.	SSG	2nd Annual (1994)	
Leon-Guerrero, Mariano R.C.	CSM	5th Annual (1997)	
Lesley, Ronald	SSG	8th Annual (2000)	
Leszczynski Jr., William J.	BG	15th Annual (2007)	
Leuer, Kenneth C.	MG	Inaugural Year (1992)	
Lincoln, Abraham	President	1st Annual (1993)	
Lindsay, James J.	GEN	2nd Annual (1994)	
Littlejohn, Carl W.	SGM	20th Annual (2012)	
Littrell, Gary L.	CSM	1st Annual (1993)	Medal of Honor
Lockett, Jr., Milton	MSG	9th Annual (2001)	
Lombardo, Roy S.	LTC	4th Annual (1996)	
Lomell, Leonard G.	1SG	2nd Annual (1994)	
Longgrear, Paul R.	COL	19th Annual (2011)	
Lucas, Andre C.	LTC	1st Annual (1993)	Medal of Honor
Lyle, James B.	COL	11th Annual (2003)	
Mace Sr., James E.	BG	15th Annual (2007)	
Madison, Eugene H.	CPT	14th Annual (2006)	
Magaña, Francisco G.	CSM	12th Annual (2004)	
Magruder III, Lawson W.	LTG	16th Annual (2008)	
Mahaffey, Fred K.	GEN	3rd Annual (1995)	
Malvesti, Richard J.	COL	20th Annual (2012)	
Mansfield, Gordon H.	HON	15th Annual (2007)	
Marion, Francis	BG	2nd Annual (1994)	
Markivich, Andy	SGM	11th Annual (2003)	
Marm, Walter J.	COL	8th Annual (2000)	Medal of Honor
Marsh, Jr, John O.	HON	1st Annual (1993)	Honorary induction
Marshall, James C.	Mr.	14th Annual (2006)	
Martin, Michael N.	CSM	3rd Annual (1995)	
Mastin, Robert L.	PFC	1st Annual (1993)	
Matos, Jr., Santos A.	SGM	7th Annual (1999)	
Matsumoto, Roy H.	MSG	1st Annual (1993)	
Mattivi, Frank	1SG	20th Annual (2012)	
McCaffrey, Barry R.	GEN	15th Annual (2007)	
McCoy, John L.	SFC	7th Annual (1999)	
McGee, George A.	COL	Inaugural Year (1992)	
McGee, John H.	BG	3rd Annual (1995)	
Mcilwain, Walter N.	LTC	11th Annual (2003)	
McLogan, Edward A.	LT	9th Annual (2001)	
Meadows, Richard J.	MAJ	4th Annual (1996)	

Inductee	Title	Year	
Meanley, Gordon W.	MAJ	16th Annual (2008)	
Merrill, Frank D.	MG	Inaugural Year (1992)	
Miles, Jr., William T.	SFC	11th Annual (2003)	
Miller, Robert J.	SSG	20th Annual (2012)	Medal of Honor
Millett, Lewis L.	COL	5th Annual (1997)	Medal of Honor
Minatra, John D.	MSG	9th Annual (2001)	
Mixon, William T.	CSM	5th Annual (1997)	
Mize, Lee	COL	4th Annual (1996)	
Moore, Frank O.	SGM	16th Annual (2008)	
Moore, Harvey L.	1SG	1st Annual (1993)	
Morrell, Glen E.	SMA	1st Annual (1993)	
Mosby, John S.	COL	Inaugural Year (1992)	
Mucci, Henry A.	COL	6th Annual (1998)	
Mueller, III, Robert S.	DIR	12th Annual (2004)	
Murphy, John F.	CPT	13th Annual (2005)	
Murphy, Robert C.	COL	20th Annual (2012)	
Murphy, Steven J.	CWO	17th Annual (2009)	
Murray, Roy A.	COL	1st Annual (1993)	
Musegades, William	MAJ	15th Annual (2007)	
Nardotti, Michael J.	MG	14th Annual (2006)	
Nett, Robert B.	COL	5th Annual (1997)	Medal of Honor
Nightingale, Keith M.	COL	19th Annual (2011)	
O'Neal, Gary L.	CWO	18th Annual (2010)	
Okamoto, Vincent H.	CPT	15th Annual (2007)	
Osborne, William Lloyd	COL	6th Annual (1998)	
Paccerelli, George A.	COL	1st Annual (1993)	
Palacios, Luis C.	CSM	19th Annual (2011)	
Palastra, Jr., Joseph T.	GEN	18th Annual (2010)	
Palmer, William T.	COL	13th Annual (2005)	
Parker, Charles H.	CPT	6th Annual (1998)	
Parker, Harris L.	CSM	10th Annual (2002)	
Peake, James B.	LTG	17th Annual (2009)	
Pentecost, Brian M.	COL	14th Annual (2006)	
Perry, Douglas M.	1SG	16th Annual (2008)	
Piazza, Philip B.	LT	3rd Annual (1995)	
Pickering, Jim R.	CSM	19th Annual (2011)	
Porter, Donn F.	SGT	1st Annual (1993)	Medal of Honor
Posey, Edward L.	MSG	10th Annual (2002)	
Powell, Colin L.	GEN	8th Annual (2000)	
Powell, Robert L.	COL	20th Annual (2012	

Inductee	Title	Year	
Prince, Robert W.	MAJ	7th Annual (1999)	
Pruden, Robert J.	SSG	Inaugural Year (1992)	Medal of Honor
Pucel, Edward W.	SFC	10th Annual (2002)	
Puckett, Ralph	COL	Inaugural Year (1992)	
Pung, Andy B.	SGT	2nd Annual (1994)	
Purdy, Donald E.	CSM	9th Annual (2001)	
Pusser, Thomas Wilson	CPT	15th Annual (2007)	
Queen, James C.	MAJ	2nd Annual (1994)	
Raaen, John C.	MG	16th Annual (2008)	
Rabel, Laszlo	SSG	Inaugural Year (1992)	Medal of Honor
Ranger, Michael B.	CPT	14th Annual (2006)	
Ray, Ronald E.	HON	9th Annual (2001)	Medal of Honor
Reed, Ellis E.	LTC	16th Annual (2008)	
Rinard, Harold L.	MSG	13th Annual (2005)	
Ripley, John W.	COL	16th Annual (2008)	Marine Corps
Rivera, Eugene C.	CPT	6th Annual (1998)	
Roberts, Marvin J.	CPT	9th Annual (2001)	
Robinson, Arthur "Robbie"	CWO	7th Annual (1999)	Honorary induction
Robison, Tom C.	SGT	13th Annual (2005)	
Rodriguez, Ramon	CSM	16th Annual (2008)	
Rogers, James D.	LTC	10th Annual (2002)	
Rogers, Robert	MAJ	Inaugural Year (1992)	
Romo, Bonifacio M.	1SG	19th Annual (2011)	
Ross, Charles G.	LTC	5th Annual (1997)	
Rudder, James E.	MG	Inaugural Year (1992)	
Salomon, Sidney A.	COL	13th Annual (2005)	
Samborowski, Leo G.	PFC	6th Annual (1998)	
Sanders, Walter M.	LTC	12th Annual (2004)	
Savory, Carlton G.	COL	17th Annual (2009)	
Schneider, Max	COL	Inaugural Year (1992)	
Scholes, Edison E.	MG	15th Annual (2007)	
Scully, Robert	SGT	16th Annual (2008)	
Shaneyfelt, Stanley E.	COL	15th Annual (2007)	
Shelton, Hugh H.	GEN	13th Annual (2005)	
Shepherd, Fred M.	SGM	3rd Annual (1995)	
Shughart, Randall D.	SFC	2nd Annual (1994)	Medal of Honor
Simons, Arthur	COL	Inaugural Year (1992)	
Singlaub, John K.	MG	14th Annual (2006)	
Singletary, Earl A.	1SG	18th Annual (2010)	
Smith, Michael H.	MSG	9th Annual (2001)	

Inductee	Title	Year	
Spears, Sam B.	CSM	11th Annual (2003)	
Spencer, Jimmie W.	CSM	12th Annual (2004)	
Spencer, Robert E.	SGM	19th Annual (2011)	
Spies, Bill	MAJ	8th Annual (2000)	
Stamper, Jr., James M.	LTC	11th Annual (2003)	
Stange III, Arthur C.	COL	1st Annual (1993)	
Stauss, Kenneth W.	LTC	10th Annual (2002)	
Stewart, Jr., Richard O.	CPT	9th Annual (2001)	
Stiner, Carl W.	GEN	12th Annual (2004)	
Storter, James G.	MAJ	12th Annual (2004)	
Strausbaugh, Leo V.	COL	16th Annual (2008)	
Stringham, Joseph S.	BG	3rd Annual (1995)	
Strong, Berkeley J.	LTC	4th Annual (1996)	
Sullivan, Eugene R.	HON	18th Annual (2010)	
Sullivan, Richard P.	LTC	5th Annual (1997)	
Sydnor, Jr., Elliot P.	COL	Inaugural Year (1992)	
Tadina, Patrick	CSM	3rd Annual (1995)	
Taylor, Wesley R.	BG	9th Annual (2001)	
Thompson, Dennis L.	SGM	10th Annual (2002)	
Thurmond, Strom	SEN	2nd Annual (1994)	Honorary induction
Toschik, Mark J.	1LT	11th Annual (2003)	
Truscott, Jr., Lucian K.	GEN	2nd Annual (1994)	Honorary induction
Turner, Robert "Tex"	COL	8th Annual (2000)	
Valeriano, Victor D.	SGT	10th Annual (2002)	
Valrey, Cleveland	CW4	13th Annual (2005)	
Van Houten, John G.	MG	14th Annual (2006)	
Versace, Humbert	CPT	8th Annual (2000)	Medal of Honor
Voorhees, Paul E.	Mr.	12th Annual (2004)	Honorary induction
Voyles, James E.	CSM	18th Annual (2010)	
Wandke, Richard D.	LTC	10th Annual (2002)	
Waters, Charles F.	CSM	14th Annual (2006)	
Watson, Martin E.	SGT	Inaugural Year (1992)	
Wawrzyniak, Stanley	LTC	19th Annual (2011)	Marine Corps
Weekley, Jr., Fredrick E.	CSM	13th Annual (2005)	
Welch, Albert C.	LTC	14th Annual (2006)	
Werner, Jr., Arthur A.	SGM	7th Annual (1999)	
Weston, Logan E.	COL	Inaugural Year (1992)	
Wijas, Rodney J.	COL	13th Annual (2005)	
Wilburn, Tom W.	1SG	13th Annual (2005)	
Wilson, Samuel V.	LTG	1st Annual (1993	

Appendix B

Rules of Discipline and Standing Orders

MAJOR ROBERT ROGERS' RULES OF DISCIPLINE

These volunteers I formed into a company by themselves, and took the more immediate command and management of them to myself; and for their benefit and instruction reduced into writing the following rules or plan or discipline, which, on various occasions, I had found by experience to be necessary and advantageous, viz.

I.   All Rangers are to be subject to the rules and articles of wars to appear at roll-call every evening on their own parade, equipped each with a firelock, sixty rounds of powder and ball, and a hatchet, at which time an officer from each company is to inspect the same, to see they are in order, so as to be ready on any emergency to march at a minute's warning; and before they are dismissed the necessary guards are to be drafted, and scouts for the next day appointed.

II.   Whenever you are ordered out to the enemy's forts or frontiers for discoveries, if your number is small, march in a single file, keeping at such a distance from each other to prevent one shot from killing two men, sending one man, or more, forward, and the like on each side, at the distance of twenty yards from the main body, if the ground you march over will admit of it, to give the signal to the officer of the approach of an enemy, and of their number etc.

III.   If you march over marshes or soft ground, change your position, and march abreast of each other, to prevent the enemy from tracking you (as they would do if you marched in a single file) till you get over such ground, and then resume your former order, and march until it is quite dark before you encamp, which do, if possible on a piece of ground that may afford your sentries the advantage of seeing or hearing the enemy at some considerable distance, keeping one half of your whole party awake alternately through the night.

IV.   Sometime before you come to the place you would reconnoiter, make a stand, and send one or two men in whom you can confide, to look out the best ground for making your observations.

V.   If you have the good fortune to take any prisoners, keep them separate till they are examined, and in your return take a different route from that in which you went out, that you may the better discover any party in your rear, and have an opportunity, if their strength be superior to yours, to alter your course, or disperse, as circumstances may require.

VI.   If you march in a large body of three or four hundred, with a design to attack the enemy, divide your party into three columns, each headed by a proper officer, and let these columns march in single files, the columns to the right and left keeping at twenty yards distance or more from that of the center, if the ground will admit, and let proper guards be kept in the front and rear, and suitable flanking parties at a due distance as before directed, with orders to halt on all eminences, to take a view of the surrounding ground, to prevent your being ambushed, and to notify the approach or retreat of the enemy, that proper dispositions may be made for attacking, defending, etc. And if the enemy approach in your front on level ground, form a front of your three columns or main body with the advanced guard, keeping out your flanking parties as if you were marching under the command of trusty officers, to prevent the enemy from pressing hard on either of your wings, or surrounding you, which is the usual method of the savages, if their number will admit of it, and be careful likewise to support and strengthen your rear guard.

VII.   If you are obliged to receive the enemy's fire, fall, or squat down, till it is over, then rise and discharge at them. If their main body is equal to yours, extend yourselves occasionally; but if superior, be careful to support and strengthen your flanking parties, to make them equal with theirs, that if possible you may repulse them to their main body, in which case push upon them with the greatest resolution, with equal force in each flank and in the center, observing to keep at a due distance from each other, and advance from tree to tree, with one half of the party before the other ten or twelve yards, if the enemy push upon you, let your front fire and fall down, and then let your rear advance through them and do the like, by which time those who before were in front will be ready to discharge again, and repeat the same alternately, as occasion shall require; by this means you will keep up such a constant fire, that the enemy will not be able easily to break your order, or gain your ground.

VIII.   If you oblige the enemy to retreat, be careful, in your pursuit of them, to keep out your flanking parties, and prevent them from gaining eminences, or rising grounds, in which case they would perhaps be able to rally and repulse in their turn.

IX.   If you are obliged to retreat let the front of your whole party fire and fall back, till the rear has done the same, making for the best ground you can; by this means you will oblige the enemy to pursue you, if they do it at all, in the face of a constant fire.

X.   If the enemy is so superior that you are in danger of being surrounded by them, let the whole body disperse, and every one take a different road to the place of rendezvous appointed for that evening, which must every morning be altered and fixed for the evening ensuing, in order to bring the whole party, or as many of them as possible, together, after any separation that may happen in the day; but if you should happen to be actually surrounded, form yourselves into a square, or if in the woods, a circle is best, and, if possible, make a stand till the darkness of the night favours your escape.

XI.   If your rear is attacked, the main body and flankers must face about to the right or left, as occasion shall require, and form themselves to oppose the enemy, as before directed, and the same method must be observed, if attacked in either of your flanks, by which means you will always make a rear of one of your flank-guards.

XII.   If you determine to rally after a retreat, in order to make a fresh stand against the enemy, by all means endeavor to do it on the most rising ground you can come at, which will give you greatly the advantage in point of situation, and enable you to repulse superior numbers.

XIII.   In general, when pushed upon by the enemy, reserve your fire till they approach very near, which will then put them into the greater surprise and consternation, and give you an opportunity of rushing upon them with your hatchets and cutlasses to the better advantage.

XIV.   When you encamp at night, fix your sentries in such a manner as not to be relieved from the main body till morning, profound secrecy and silence being often of the last importance in these cases. Each sentry, therefore, should consist of six men, two of whom must be constantly alert, and when relieved by their fellows, it should be done without noise; and in case those on duty see or hear anything, which alarms them, they are not to speak, but one of them is silently to retreat, and acquaint the commanding officer thereof, that proper dispositions may be made and ail occasional sentries should be fixed in like manner.

XV.   At the first dawn of day, awaken your whole detachment; that being the time when the savages choose to fall upon their enemies, you should by all means be in readiness to receive them.

XVI.   If the enemy should be discovered by your detachments in the morning and their numbers are superior to yours, and a victory doubtful, you should not attack them till the evening, as then they will not know your numbers, and if you are repulsed, your retreat will be favoured by the darkness of the night.

XVII.   Before you leave your encampment, send out small parties to scout round it, to see if there be an appearance or track of an enemy that might have been near you during the night.

XVIII.   When you stop for refreshments, choose some spring or rivulet if you can, and dispose your party so as not to be surprised posting proper guards and sentries at a due distance, and let a small party waylay the path you came in, lest the enemy should be pursuing.

XIX.   If, in your return, you have to cross rivers, avoid the usual fords as much as possible, lest the enemy should have discovered, and be there expecting you.

XX.   If you have to pass by lakes, keep at some distance from the edge of the water, lest, in case of an ambuscade, or an attack from the enemy, when in that situation, your retreat should be cut off.

XXI.   If the enemy pursue your rear, take a circle till you come to your own tracks, and there form an ambush to receive them, and give them the first fire numbers.

XXII.   When you return from a scout, and come near our forts, avoid the usual roads, and avenues thereto, lest the enemy should have headed you, and lay in ambush to receive you, when almost exhausted with fatigue.

XXIII.   When pursue any party that has been near our forts or encampments, follow not directly in their tracks, lest you should be discovered by their rear guards, who, at such a time, would be most alert; but endeavor, by a different route, to head and meet them in some narrow pass or lay in ambush to receive them when and where they least expect it.

XXIV.   If you are to embark in canoes, bateaux, or otherwise by water, choose the evening for the time of your embarkation, as you will then have the whole night before you, to pass undiscovered by any parties of the enemy, on hills, or other places, which command a prospect of the lake or river you are upon.

XXV.   In paddling or rowing, give orders that the boat or canoe next the sternmost, wait for her, and the third for the second, and the fourth for the third, and so on, to prevent separation, and that you may be ready to assist each other on any emergency.

XXVI.   Appoint one man in each boat to look out for fires, on the adjacent shores, from the numbers and size of which you may form some judgment of the number that kindled them, and whether you are able to attack them or not.

XXVII.   If you find the enemy encamped near the banks of a river, or lake, which you imagine they will attempt to cross for their security upon being attacked, leave a detachment of your party on the opposite shore to receive them, while, with the remainder, you surprise them, having them between you and the lake or river.

XXVIII.   If you cannot satisfy yourself as to the enemy's number and strength, from their fire, etc., conceal your boats at some distance, and ascertain their number by a reconnoitering party, When they embark, or march, in the morning, marking the course they steer, etc., when you may pursue, ambush, and attack them, or let them pass, as prudence shall direct you. In general, however, that you may not be discovered by the enemy on the lakes and rivers at a great distance, it is safest to lay by, with your boats and party concealed all day, without noise or show, and to pursue your intended route by night; and whether you go by land or water, give out parole and countersigns, in order to know one another in the dark, and likewise appoint a station for every man to repair to, in case of any accident that may separate you.[1]

ROGERS' STANDING ORDERS

1.   Don't forget nothing.
2.   Have your musket clean as a whistle, hatchet scoured, sixty rounds powder and ball, and be ready to march at a minute's warning.
3.   When you're on the march, act the way you would if you was sneaking up on a deer. See the enemy first.
4.   Tell the truth about what you see and what you do. There is an Army depending on us for correct information. You can lie all you please when you tell other folks about the Rangers, but don't ever lie to a Ranger or officer.
5.   Don't never take a chance you don't have to.
6.   When you're on the march we march single file, far enough apart so one shot can't go through two men.
7.   If we strike swamps, or soft ground, we spread out abreast, so it's hard to track us.
8.   When we march, we keep moving till dark, so as to give the enemy the least possible chance at us.
9.   When we camp, half the party stays awake while the other half sleeps.

10. If we take prisoners, we keep 'em separate till we have time to examine them, so they can't cook up a story between 'em.

11. Don't ever march home the same way. Take a different route so you won't be ambushed.

12. No matter whether we travel in big parties or little ones, each party has to keep a scout twenty yards on each flank and twenty yards in the rear, so the main body can't be surprised and wiped out.

13. Every night you'll be told where to meet if surrounded by a superior force.

14. Don't sit down to eat without posting sentries.

15. Don't sleep beyond dawn. Dawn's when the French and Indians attack.

16. Don't cross a river by a regular ford.

17. If somebody's trailing you, make a circle, come back onto your tracks, and ambush the folks that aim to ambush you.

18. Don't stand up when the enemy's coming against you. Kneel down, lie down, hide behind a tree.

19. Let the enemy come till he's almost close enough to touch. Then let him have it and jump out and finish him up with your hatchet.[2]

Notes

INTRODUCTION

1. At the time of this writing the United Arab Emirates (UAE) Land Forces were under a complete transformation effort where the Crown Prince has hired retired US Army officers under the tutelage of Lieutenant General Thomas T. Turner (a combat experienced Ranger earning his combat scroll in the Panama Invasion in 1989 and former Commander of 5th Army, Commander of the Southern European Task Force or SETAF and the Commander of the 101st Airborne Division) to change their forces into a more combat effective fighting force in the Middle East.

2. Immediately after their failure in Indochina, the French Supreme Command, Far East, prepared a candid appraisal of their performance. The results were published in 1955 in three volumes, summarizing lessons dealing with politico-military concerns, counterinsurgency in general, and tactics. General Paul Ely, Commander-in-Chief, Indochina, described his study as a collective self-appraisal. He intoned we must review the causes of our failures and of our successes to ensure that the lessons which we bought so dearly with our dead not remain locked away in the memories of the survivors. He defended his candor by arguing an Army with a long history is sufficiently well-endowed to be able to hear the truth. Based on 1,400 reports written by officers of all ranks, the volumes dwell heavily on the need for pacification and political action at the village level, properly coordinated with all military actions. Assessments were indeed candid, and provided clear lessons for the French Army to apply in future counterinsurgencies such as in Algeria. The French clearly aimed to learn from their mistakes and to do better next time. And they did. The fact that the nation eventually lost the war in Algeria should not detract from the fact that the French Army's counterinsurgency performance was much improved. This example would seem to support the common maxim that military defeat is an army's best teacher, as it eliminates incompetent leaders and practices, promotes innovative reforms, and forces deficiencies to be fixed. However, historian Edward Drea noted that such generalizations overstate the case. He writes, the way an army interprets defeat in relation to its military tradition, and not the defeat itself, will determine, in large measure, the impact an unsuccessful military campaign will have on that institution. Edward J. Drea, "Tradition and Circumstance: The Imperial Japanese Army's Tactical Response to Khalkin-Gol, 1939," in Colonel Charles R. Shrader, ed., *Proceedings of the 1982 International Military History Symposium: The Impact of Unsuccessful Military Campaigns on Military Institutions, 1860-1980*, Washington, DC: US Army Center of Military History, 1984, p. 134.

3. The United States discontinued the draft in 1973, moving to an all-volunteer military force resulting in the disbanding of a mandatory conscription. However, the Selective Service System remains in place as a contingency plan where men between the ages of 18 and 25 are required to register so that a draft can be readily resumed if needed. As of this writing the strategic conditions exists where conscription is considered unlikely by most political and military experts. In fact, some experts consider the Selective Service System to be pointless since the chances of a draft are nearly zero since it's a common understanding that the US populace, as of this writing, would be unwilling to support a draft. In fact, as the wars in Iraq and Afghanistan ensued for the past decade-plus very little of the nation besides portions of the military was engaged in any way to support the cause. There is no doubt that the current national conditions would not support a resurgence of a draft. In colonial times, the Thirteen Colonies used a militia system for local defense. Colonial militia laws—and after independence those of the United States and the various states—required able-bodied males to enroll in the militia, to undergo a minimum of military training and to serve for limited periods of time in war or emergency. This earliest form of

conscription involved selective drafts of militiamen for service in particular campaigns. Following this system in its essentials, the Continental Congress in 1778 recommended that the states draft men from their militias for one year's service in the Continental army. This first national conscription was irregularly applied and failed to fill the Continental ranks. For long-term operations, conscription was occasionally used when volunteers or paid substitutes were insufficient to raise the needed manpower. During the American Revolution, the states sometimes drafted men for militia duty or to fill state Continental Army units, but the central government did not have the authority to conscript. President James Madison and his Secretary of War James Monroe unsuccessfully attempted to create a national draft of 40,000 men during the War of 1812. This proposal was fiercely criticized on the Senate floor by legendary Massachusetts Senator Daniel Webster in "one of [his] most eloquent efforts." Webster, Daniel (1814-12-09) "On Conscription," reprinted in *Left and Right: A Journal of Libertarian Thought* (Autumn 1965).

4.   Thomas W. Evans (Summer 1993). "The All-Volunteer Army after Twenty Years: Recruiting in the Modern Era". Sam Houston State University found at http://www.shsu.edu/~his_ncp/VolArm. html accessed May 5, 2009.

5.   Jonathan Aitken, *Nixon: A Life*, Washington DC: Public Affairs, Regency Publishing, Inc., 2008, pp. 396–397.

6.   Senatorial opponents of the war wanted to reduce this to a one-year extension, or eliminate the draft immediately altogether, or tie the draft renewal to a timetable for troop withdrawal from Vietnam. Senator Mike Gravel of Alaska took the most forceful approach trying to filibuster the draft renewal legislation, shut down conscription, and directly force an end to the war. Senators supporting Nixon's war efforts supported the bill, even though some had qualms about ending the draft. In September 1971 after a prolonged battle in the Senate, closure was achieved over the filibuster and the draft renewal bill was approved. Jonathan Aitken, *Nixon: A Life*, Washington DC: Public Affairs, Regency Publishing, Inc., 2008, pp. 396–397 and http://www.rand.org/pubs/research_briefs/RB9195/index1.html accessed on October 28, 2011.

7.   Aitken, *Nixon: A Life*, pp. 396–397.

8.   Robert K. Griffith, Jr. and Wyndham Mountcastle, John, *U.S. Army's Transition to the All-volunteer Force, 1868-1974*, Darby, Pennsylvania: Diane Publishing, 1997, pp. 40–41.

9.   http://www.sss.gov/lotter1.htm accessed on July 6, 2014.

10. Arlington National Cemetery Website, "Creighton Williams Abrams, Jr.", found at http://www.arlingtoncemetery.com/abrams.htm accessed May 12, 2012.

11. Lewis Sorley, "The Art of Taking Charge," *Across the Board,* May 1992, p. 35.

12. Arlington National Cemetery Website, "Creighton Williams Abrams, Jr." found at http://www.arlingtoncemetery.com/abrams.html access on May 13, 2012.

13. Kent T. Woods, "Rangers Lead the Way: The Vision of General Creighton W. Abrams," US Army War College Strategy Research Project, Carlisle Barracks, Pennsylvania, July, 14, 2003, pp.2-3.

CHAPTER 1

1   Lock, *To Fight with Intrepidity*, p. 1.

2   Ross Hall, *The Ranger Book: A History 1634-2006*, 2007, p. 9.

3   Lock, *To Fight with Intrepidity*, p. xvi.

4   Philip L. Barbour, ed., *The Complete Works of Captain John Smith,* Chapel Hill, North Carolina: University of North Carolina Press, 1986, p. 264.

5   Robert W., Black, *Ranger Dawn: the American Ranger from the Colonial Era to the Mexican War*, Mechanicsburg, Pennsylvania: Stackpole Books, 2009. p. 4.

6   Black, p. 4.

7   *Indian Narratives* (Claremont, New Hampshire: Tracy and Brothers, 1854), pp. 262, 264

8   In 1371 Henry Dolyng in his forest districts was a Ranger of the New Forest and Thomas of Croydon was a Ranger in Waltham. Mir Banmanyar's http://www.suasponte.com/earlyrang.htm accessed on September 1, 2012.

9   William Claiborne (1600-1677) was an English pioneer, surveyor, and early settler of Virginia and Maryland. Becoming a wealthy planter, trader and politician he was a central figure in the disputes

involving Kent Island in the Chesapeake Bay which eventually evolved into the first naval battle in North American waters.

10 King Philip's War, sometimes called called the First Indian War, Metacom's War, Metacomet's War, or Metacom's Rebellion, was an armed conflict between Native American inhabitants of present-day southern New England and English colonists and their Native American allies in 1675–76. The war is named after the main leader of the Native American side, Metacomet, known to the English as "King Philip." Major Benjamin Church emerged as the Puritan hero of the war; it was his company of Puritan Rangers and Native American allies that finally hunted down and killed King Philip on August 12, 1676. The war continued in northern New England (primarily on the Maine frontier) after King Philip was killed, until a treaty was signed at Casco Bay in April 1678. Many define this war as the bloodiest war in America's history on a per capita basis, which the Rangers won.

11 Hall, *The Ranger Book,* p. 10.

12 Black, *Ranger Dawn,* p. 21.

13 Praying Indian is a 17th century term referring to Native Americans of New England who converted to Christianity. While many groups are referred to by this term, it is more commonly used for tribes that were organized into villages, known as praying towns by Puritan leader John Eliot. http://en.wikipedia.org/wiki/Benjamin_Church_ (military_officer), accessed on September 1, 2012.

14 The Acadians are the descendants of the 17th-century French colonists who settled in Acadia (located in the Canadian Maritime provinces – Nova Scotia, New Brunswick, and Prince Edward Island – as well as part of Quebec, and in the US state of Maine). Acadia was a colony of New France. Although today most of the Acadians and Québécois are French speaking Canadians, Acadia was a distinct colony of New France, and was geographically and administratively separate from the French colony of Canada (modern day Quebec), which led to Acadians and Québécois developing two rather distinct histories and cultures. The settlers whose descendants became Acadians came from "all the regions of France but coming predominantly directly from the cities." Prior to the British Conquest of Acadia in 1710, the Acadians lived for almost 80 years in Acadia. After the Conquest, they lived under British rule for the next 45 years. During the French and Indian War, British colonial officers and New England legislators and militia carried out the Great Expulsion of 1755–1763. They deported approximately 11,500 Acadians from the maritime region. Approximately one-third perished from disease and drowning. One historian compared this event to a contemporary ethnic cleansing, while other historians suggested that the event is comparable with other deportations in history. Many later settled in Louisiana, where they became known as Cajuns. Others were transported to France. Later on many Acadians returned to the Maritime Provinces of Canada, most specifically New Brunswick. Most who returned ended up in New Brunswick because they were barred by the British from resettling their lands and villages in the land that became Nova Scotia. This was a British policy to assimilate them with the local populations where they resettled. Acadians speak a dialect of French called Acadian French. Many of those in the Moncton, New Brunswick area speak Chiac and English. The Louisiana Cajun descendants mostly speak English, with a distinct local dialect known as Cajun English being prominent, but some still speak Cajun French, a French dialect developed in Louisiana. http://en.wikipedia.org/wiki/Acadians accessed on July 1, 2012.

15 Hall, *The Ranger Book,* p. 11.

16 *Northwest Passage* is a 734 page historical novel by Kenneth Roberts, published by Doubleday Books in 1937. Told through the eyes of primary character Langdon Towne, much of this novel centers on the exploits and character of Robert Rogers, the leader of Rogers' Rangers, who were a colonial force fighting with the British during the French and Indian War, Lock, *To Fight with Intrepidity,* p. 9.

17 Upon assignment to the Regiment, both officers and senior NCOs attend the Ranger Orientation Program (ROP) to integrate them into the Regiment. ROP familiarizes them with Regimental policies, standing operating procedures, the Commander's intent and Ranger standards. Enlisted soldiers assigned to the Regiment go through the Ranger Indoctrination Program (RIP). RIP assesses potential Rangers on their physical qualifications and indoctrinates basic Regimental standards. Soldiers must pass ROP or RIP to be assigned to the 75th Ranger Regiment, multiple sources to include Hall, *The Ranger Book* and Lock, *To Fight with Intrepidity.*

18 This paradox; of irregular fighters functioning within the hierarchical structure of a regular army, would continue to trouble future Ranger units until the lines became more blurred in the recent counterinsurgency forays in Iraq and Afghanistan. During both Operation Iraqi Freedom and Enduring Freedom the Charter is more obvious than ever for the operating environment seen in both of these wars forced conventional forces to have inherited many of the tasks, techniques and procedures (TTPs) historically displayed by the special operation forces like the US Army Rangers. Mir Banmanyar's http://www.suasponte.com/earlyrang.htm accessed on September 2, 2012.

19 American Archive Documents of the American Revolution, 1774-1776, produced by Northern Illinois University Library p. V2:1847 found at http://lincoln.lib.niu.edu/cgi-bin/amarch/getdoc. pl?/var/lib/philogic/databases/amarch/.5774 accessed on February 2, 2012.

20 *Defense Intelligence Agency* homepage at http://www.dia.mil/News/Articles/tabid/3092/ Article/8826/knowltons-rangers-but-one-life-to-give.aspx accessed on May 27, 2014.

21 Military Intelligence Corps Association website at http://www.micastore.com/Awards.html accessed January 15, 2012.

22 http://www.fideles.net/fidelesrangers.cfm accessed on January 17, 2012.

23 http://www.smithsonianmag.com/history-archaeology/biography/fox.html#ixzz1jpKyjzwW accessed on January 18, 2012.

24 http://www.patriotshistoryusa.com/teaching-materials/bonus-materials/american-heroes-francis-marion accessed on July 1, 2012.

25 http://www.smithsonianmag.com/history-archaeology/biography/fox.html#ixzz1jjkR8oOd, accessed on January 16, 2012

26 Hall, p. 38.

27 Lock, *To Fight with Intrepidity,* p. 193.

28 http://www.fideles.net/about/fideles-rangers.cfm accessed on September 3, 2012 and Jung, *The Black Hawk War of 1832,* pp. 79-85.

29 Charles F. Ritter and Jon L Wakelyn, *Leaders of the American Civil War A Biographical and Historiographical Dictionary,* Greenwood Publishing Group, 1998, p. 236.

30 Albert Castel, *Civil War Kansas: Reaping the Whirlwind,* Lawrence, Kansas: University Press of Kansas 1997, pp. 1-2.

31 The Confederacy declared that these partisan Rangers would be regarded as outlaws and that only state sponsored, government authorized, go-by-the-rules units would be allowed. To ensure of it they developed the Partisan Ranger Act of April 21, 1862 that authorized an organized establishment of Ranger units comprised of specifically authorized commanders. This act directed that these established units be armed and paid by the Confederate Government and that the Rules of War applied to them in their actions. Jeremy B. Miller, "Unconventional Warfare in the American Civil War," Fort Leavenworth, Kansas: Command and General Staff College, 2004, p. 14.

32 Hall, p. 41.

33 *Ranger Handbook (SH 21-76)* found at http://www.africom.mil/WO-NCO/DownloadCenter/%5 C40Publications/Ranger%20Handbook.pdf accessed on December 31, 2011.

34 Jeffry D. Wert, *Cavalryman of the Lost Cause: A Biography of J. E. B. Stuart,* p. 207.

35 http://www.corydonbattlepark.com/battle.html access on July 1, 2012.

36 John Esten Cooke, *Wearing of the Gray,* Whitefish, Montana: Kessinger Publishing, LLC, 2010, p. 80.

37 Lock, *Rangers in Combat: A Legacy in Valor,* P. 85.

38 Hall, p. 67.

39 Of note, Murray was six months older than his commander William O. Darby. Robert W. Black, in his book *Rangers in World War II,* makes these observations about Roy Murray. "Murray was an athlete and outdoorsman, a cross-country runner whose hobbies included hiking and fishing. He had experience in civilian life in navigation and boat handling. Murray had keen analytical and good communication skills. His recommendations were sound, and he was a strong leader who had a profound influence on subsequent Ranger activities." James Altieri wrote of Murray, "The afternoon of the first day was entirely devoted to sports. Some fellows played baseball, some wrestled, some put on boxing gloves, while the rest played football. I joined the group playing football and got a good opportunity to study the slim, trim captain who was commanding our company.

He was as fast as a deer and as tough as nails. I guessed him about thirty-years old, but he moved with the agility of a nineteen-year-old. What impressed me most was that he made no effort to dominate the team's strategy. His name was Roy Murray." http://www.irelandseye.com/aarticles/history/events/worldwar/rangers2.shtm accessed on July 1, 2012.

40 http://www.americanmilitaryhistorymsw.com/blog/620296-us-army-rangers-at-dieppe/ accessed on January 22, 2012.
41 Lock, *To Fight with Intrepidity*, p. 253.
42 A Ranger saying goes that the Ranger Tab is an acquired skill set and a badge; the *scroll* is a way of life.
43 http://armyranger.com/index.php/history/modern-era accessed on September 5, 2012.
44 "Darby's Rangers" found at http://www.rangerfamily.org/Commanders/Wm%20O%20Darby.htm accessed on July 7, 2014.
45 Mir Banmanyar's http://www.suasponte.com/WWII.htm accessed on September 5, 2012.
46 Ibid.
47 Robert J. Martin and Thomas H. Taylor, *Rangers Lead the Way*, Nashville, Tennessee: Turner Publishing, 1996, p. 54.
48 Mir Banmanyar's http://www.suasponte.com/WWII.htm accessed on September 5, 2012.
49 Stephen E. Ambrose, *D-Day: June 6, 1944, The Climatic Battle of World War II*, New York: Simon & Schuster, 1995, p. 430.
50 Mir Banmanyar's http://www.suasponte.com/WWII.htm accessed on September 5, 2012.
51 Robert W. Black, *Rangers in World War II*, New York, New York: Ballentine Books, Random House, 1992, p. 206.
52 Mona D. Sizer, *The Glory Guys: The Story of the US Army Ranger*, Lanham, Maryland: Taylor Trade Publishing 2005, p. 224.
53 Ibid.
54 Mir Banmanyar's http://www.suasponte.com/WWII.htm accessed on September 7, 2012.
55 Ambrose, *D-Day*, p. 431.
56 http://armyranger.com/index.php/history/modern-era accessed on September 7, 2012.
57 Interview with Brigadier General (retired) William J. Leszczynski, Jr, the 9th Colonel of the 75th Ranger Regiment about the heights of the cliffs that he attained while serving on the American Battles Monument Commission. Interview on December 10, 2011 and John A. Adams and Henry C. Dethloff, *Texas Aggies go to War: In Service of Their Country, College Station*, Texas: Texas A & M University Press, 2008, p. 160.
58 http://www.digplanet.com/wiki/Pointe_du_Hoc accessed on September 7, 2012.
59 http://www.rangerfamily.org/History/History/Battalion%20Pages/sixth.htm accessed on September 7, 2012.
60 http://armyranger.com/index.php/history/modern-era accessed on September 8, 2012.
61 High Speed Transports were converted destroyers and destroyer escorts used to support amphibious operations in World War II and afterward. They received the US Hull classification symbol APD; "AP" for transport and D for destroyer. APDs were intended to deliver small units such as Marine Raiders, Underwater Demolition Teams and United States Army Rangers onto hostile shores. They could carry up to a company size unit. They also provided gunfire support as needed. http://citizendia.org/6th_Ranger_Battalion (United States) accessed on September 8, 2012.
62 Contributed by Leo V. Strausbaugh (Colonel, US Army retired) at http://www.wwiirangers.com/History/History/Battalion%20Pages/sixth.htm accessed on January 31, 2012.
63 Lock, *To Fight with Intrepidity*, p. 307.
64 Mir Banmanyar's http://www.suasponte.com/korea.htm accessed on September 8, 2012.
65 Hall, p. 244.
66 Fort Benning is named for Brigadier General Henry L. Benning, a Confederate army general and a native of Columbus, Georgia. It was established in 1918 as Camp Benning.
67 The reader may find it interesting to refer to *The Early Years* portion of this chapter to draw parallels to the need for Rangers to counter the Indian type LRRP/LRP tactics seen in the French and Indian Wars. http://www.armyranger.com/index.php/history accessed on September 8, 2012.
68 Multiple sources to include Mir Banmanyar's http://www.suasponte.com/vietnam.htm accessed on July 1, 2012, September 8, 2012 and May 29, 2014.

69 With one notable WWII exception, since 1816, US Army regiments have not included a Juliet or "J" company. Shelby L. Stanton, Vietnam Order of Battle, Mechanicsburg, Pennsylvania: Stackpole Books, 2003, p. 154.

70 The tradition of wearing black berets began with armored units. In 1924 the British Royal Tank Regiment adopted the first modern military beret, based on the Scottish highland bonnet and French Bretonne beret. The regiment selected the headgear for its practicality--brimless for use with armored vehicle fire control sights and black to hide grease stains. In the US Army, HQDA policy from 1973 through 1979 permitted local commanders to encourage morale-enhancing distinctions, and Armor and Armored Cavalry personnel wore black berets as distinctive headgear until Chief of Staff of the Army Bernard W. Rogers banned all such unofficial headgear in 1979. Rangers received authorization through AR 670-5, Uniform and Insignia, January 30, 1975, to wear black berets. Previously, locally authorized black berets had been worn briefly by the 10th Ranger Company (Airborne), 45th Infantry Division, during the Korean War before their movement to Korea; Company F (LRP), 52d Infantry, 1st Infantry Division, in 1967 in the Republic of Vietnam; Company H (Ranger), 75th Infantry, 1st Cavalry Division, in 1970 in the Republic of Vietnam; and Company N (Ranger), 75th Infantry, 173rd Airborne Brigade, in 1971 in the Republic of Vietnam. In June, 2001 the Chief of Staff of the Army Eric Shinseki authorized the black beret for the entire Army. The Rangers complied and changed their beret to a tan color. Found at http://www.army.mil/features/beret/beret.html, accessed on February 8, 2012. When the Ranger beret was switched from the black to the tan beret the Regimental Commander at the time, P.K. Keen issued the following statement: Fellow Rangers, The purpose in writing this note is to inform you that the 75th Ranger Regiment will exchange our traditional Black Beret for a Tan Beret. The Army's donning of the Black Beret, as its standard headgear is a symbol of the "Army's on-going Transformation" and a "symbol of excellence." The 75th Ranger Regiment fully supports our Army's initiative to don the Black Beret. The Tan color of the new Ranger Beret reinvigorates the historical and spiritual linkage throughout the history of the American Ranger. It is the color of the buckskin uniforms and animal skin hats of Rogers' Rangers, the first significant Ranger unit to fight on the American continent, and the genesis of the American Ranger lineage. Tan is the one universal and unifying color that transcends all Ranger Operations. It reflects the Butternut uniforms of Mosby's Rangers during the American Civil War. It is reminiscent of the numerous beach assaults in the European Theater and the jungle fighting in the Pacific Theaters of World War II, where Rangers and Marauders spearheaded victory. It represents the khaki uniform worn by our Korean and Vietnam War era Rangers and the color of the sand of Grenada, Panama, Iraq, and Mogadishu, where modern day Rangers lead the way as they fought and, at times, valiantly died accomplishing the Ranger mission. Tan rekindles the legacy of Rangers from all eras and exemplifies the unique skills and special capabilities required of past, present, and future Rangers. The Ranger Tan Beret will distinguish Rangers in the 21st Century as the Black Beret recognized them as a cut above in the past. With the donning of this new Beret, rest assured that the 75th Ranger Regiment will continue to Lead the Way with its high standards. I made this decision because I feel it is best for the Ranger Regiment and our Army, today and in the future. Following the announcement that on 14 June 2001 the Army would adopt the Black Beret as its standard headgear I asked the Regimental Command Sergeant Major to put together a uniform committee to examine some possible uniform options for the Regiment. These options included maintaining the current Black Beret, adding distinctive insignia to the Black Beret, and adopting a different color beret (ultimately six different colors were examined). The committee I established met three times over two months to consider input from Rangers of all ranks in the Regiment. The members of this group included the Honorary Colonel of the Regiment, DCO (Deputy Commanding Officer), RSM (Regimental Sergeant-Major), CSMs (Command Sergeant-Major) of each Battalion, and 1SGs of RHHC (Regimental Headquarters and Headquarters Company) and RTD (Ranger Training Detachment). From the initial options, the committee narrowed consideration to maintaining the current Black Beret, augmenting the Black Beret with a WWII Ranger "diamond" patch attached next to the flash, and an option of replacing the Black Beret with a Tan colored beret. The committee explored each option historically giving equal consideration to its appearance when donned with each of our uniforms. After receiving input from the units, the Tan

Beret was selected. Shortly after 1st Ranger Battalion was reactivated in 1974, the Army formally authorized the Black Beret for Rangers. By so doing, I do not believe it was saying the Rangers were different from the rest of the Army, but that they were distinctive within the Army, that more was expected of them, and that they would set the standards for the rest of the Army. They would be asked to "Lead the Way" as Rangers had done since WWII. As today's Rangers follow in the footsteps of those who preceded them, they continue to uphold the high standards of the Regiment as they prepare for tomorrow's battles. Changing from the Black Beret to the Tan Beret is not about being different from the rest of the Army, but about a critical aspect that unifies our Army and makes it the best Army in the world – High Standards. One of the Rangers' most visible distinctive "physical features" is the beret. In the past, the beret distinguished the Rangers and acknowledged that they were expected to maintain higher standards, move further, faster, and fight harder than any other soldiers. I believe Rangers today and in the years to come deserve that same distinction. Rangers have never been measured by what they have worn in peace or combat, but by commitment, dedication, physical and mental toughness, and willingness to Lead the Way – Anywhere, Anytime. The Beret has become one of our most visible symbols, it will remain so. Unity within our Army is absolutely critical to combat readiness and Rangers have always prided themselves in being part of that unity. Unity among Rangers, past and present, is essential to moving forward and ensuring we honor those who have put the combat streamers on our colors and acknowledge the sacrifices and dedication of the Rangers and their families who serve our nation today. I hope that when our Army dons the Black Beret and our Rangers put on the Tan Beret we will move forward and focus on what is ultimately the most important task in front of us – ensuring the continued high state of Readiness of the Ranger Regiment. We can do that by training hard and taking care of our Rangers and their families. The continued support of all Rangers to our Army is important to sustaining that Readiness. Thanks to our Army, the 75th Ranger Regiment today is fully resourced and combat ready. Our focus in the future is maintaining that high state of readiness. Again, thanks to each of you for everything you have done for our nation and our Rangers. Rangers Lead the Way! P. K. Keen Colonel, Infantry 11th Colonel of the Regiment. Mir Bahmanyar, *Shadow Warriors: A History of the US Army Rangers*, New York, New York: Random House Inc., 2005, pp. 168-169.

71 *Lineage and Honors, 75th Ranger Regiment*, Washington DC: The Department of the Army, March 26, 2013 found at http://www.history.army.mil/html/forcestruc/lineages/branches/inf/0075ra.htm accessed on July 7, 2014.

CHAPTER 2

1. http://www.history.army.mil/books/AMH-V2/AMH%20V2/chapter12.htm, accessed on September 9, 2012.
2. Kent T. Woods, "Rangers Lead the Way: The Vision of General Creighton W. Abrams," U.S. Army War College Strategy Research Project, Carlisle Barracks, PA, July, 14, 2003, pp. 3–4.
3. Ibid, p. 4.
4. VOLAR refers to the Volunteer Army, General Creighton W. Abrams, notes from Staff Meeting, February 26, 1973, Military History Institute, Carlisle Barracks, Pennsylvania.
5. Kim Wellenson, "Drugs 2-22," *Wire Services News*, Washington, DC, February 22, 1973.
6. http://www.history.army.mil/books/AMH-V2/AMH%20V2/chapter12.htm, accessed on September 24, 2012.
7. http://www.marxists.de/war/geier/vietnam.htm, accessed on September 9, 2012.
8. Matthew Rinaldi, "The Olive-Drab Rebels: Military Organizing during the Vietnam Era," *Radical America*.8 no. 3, May–June 1974, 29.
9. Richard A. Gabriel and Paul L. Savage, *Soldiers in Revolt: The American Military Today* (Garden City, NY: Doubleday, 1975), 35–36.
   David Cortright, *Crisis in Command: Mismanagement in the Army*, New York: Hill and Wang, 1978, 254.
10. http://www.democraticunderground.com/discuss/duboard.php?az=view_all&address=104x55109, accessed on September 9, 2012.

11. David Cortright, *Soldiers in Revolt: The American Military Today* (Garden City, NY: Doubleday, 1975), 35–36.
12. http://www.marxists.de/war/geier/vietnam.htm, accessed on September 9, 2012.
13. David Zirin, *People's History of Sports in the United States: 250 Years of Politics, Protest, People, and Play* New York, New York: New Press, 2009, 182.
14. http://www.democraticunderground.com/discuss/duboard.php?az=view_all&address=104x55109, accessed on September 9, 2012.
15. Joel Geier, "Vietnam: The Soldier's Revolt," *International Socialist Review* no. 9, August-September 2000, http://www.isreview.org/issues/09/soldiers_revolt.shtml, accessed on March 3, 2012.
16. http://www.biographyonline.net./sport/muhammad_ali.html, accessed on March 3, 2012
17. Robert D. Heinl Jr., "The Collapse of the Armed Forces," *Armed Forces Journal* (June 7, 1971), 1–2, http://msuweb.montclair.edu/~furrg/Vietnam/heinl.html, accessed on February 9, 2012.
18. G. S. Eckhardt, preface to the "Study on Military Professionalism," US Army War College, Carlisle Barracks, PA, June 30, 1970.
19. U.S. Army War College, "Study on Military Professionalism," Carlisle Barracks, PA, June 30, 1970, p. B-1-9.
20. Ibid, pp. 48 and B-1-9.
21. Ibid, 31.
22. Richard W. Stewart, ed., *American Military History: The United States Army in a Global War, 1917–2003, Volume II*, Washington, DC: Center of Military History, "Rebuilding the Army: Vietnam to Desert Storm," 2005, found at http://www.history.army.mil/books/AMH-V2/AMH%20V2/chapter12.htm, accessed on February 15, 2012, 369.
23. Eric Foner, *Reconstruction: America's Unfinished Revolution, 1863–1877*, The New American Nation series, New York: New York: Harper & Row, 2008, p. 32.
24. Today's noncommissioned officer education system (NCOES) is the keystone for NCO development. It provides leader and MOS skill training in an integrated system of resident training at four levels (primary, basic, advanced, and senior). Warrior Leader Course (WLC), formerly known as PLDC (Primary Leader Development Course), is the first school in the NCOES. The Advanced Leader Course (ALC), formerly BNCOC (Basic Noncommissioned Officers Course), is typically attended by NCOs in the ranks of sergeant and staff sergeant. The Senior Leader Course (SLC) is a new name for the Advanced Noncommissioned Officers Course (ANCOC) and is generally attended by NCOs of the rank of staff sergeant and sergeant first class. The First Sergeant Course is for NCOs who will be promoted to the rank of first sergeant, and the Sergeant Major Course is the top course in the NCOES and a resident course held at Fort Bliss, Texas, at the U.S. Army Sergeant Major Academy.
25. Interviews with multiple retired officers and NCOs were used for the firsthand thoughts and experiences found in this and other chapters of this book. Unless it is a direct quote, acknowledgment of each interview is not cited in order to reduce the number of redundant endnote entries. A list of interviews can be found in the bibliography.

CHAPTER 3

1. Hall, *The Ranger Book*, p. 441.
2. Gorman who preceded Richardson's at Fort Benning as the Assistant Commander of the Infantry center and would eventually command the US Southern Command from 1983-1985.
3. Youssef Aboul-Enein, Infantry Magazine, September -October 2010 found at http://findarticles.com/p/articles/mi_m0IAV/is_3_99/ai_n56541299 accessed on March 5, 2012.
4. Jeffrey S. Wilson, "Transformational Leadership: William DePuy's Vision for the Army," *Military Review* (September 1, 2011), found at http://readperiodicals.com/201109/248091260.html accessed on March 5, 2012.
5. DePuy, William, *Changing an Army*, Washington, DC: Center of Military History, 1988 found at http://www.history.army.mil/html/books/070/70-23/CMH_Pub_70-23.pdf, p. 193 accessed on February 1, 2015.
6. http://schuylkillcountymilitaryhistory.blogspot.com/2011/01/greetings-draft-notice-from-civil-war.html accessed pm September 10, 2012.

7. Hall, *The Ranger Book,* p. 445.
8. Ibid.
9. Ibid.
10. Leuer would go on to retire as a Major General and command Fort Benning at the end of his career; an ironic position to grow into given he was so dependent upon Richardson and the Fort Benning community in his "stand-up" of the 1st Ranger Battalion.
11. Lock, *To Fight with Intrepidity,* p. 439.
12. Francis H. Kearney III, "The impact of leaders on organizational culture: a 75th ranger regiment case study," Carlisle Barracks, Pennsylvania: US Army War College, 1997, p. 12.
13. Commander, Military Personnel Center, Subject: Application for Assignment to Ranger Duty. AUTODIN Message to AIG 7401, Alexandria, Virginia, March 5, 1973.
14. Ken Keen, "The Ranger Regiment: Strategic Force for the 21st Century: Strategy Research Project, Carlisle Barracks, Pennsylvania: US Army War College, April 1, 1998, p. 5.
15. Lock, *To Fight with Intrepidity,* p. 440.
16. Hall, *The Ranger Book,* pp. 429-430.
17. John Baker, "SUBJECT: Personnel Actions," Memorandum to the Commander, XVIII Airborne Corps and Fort Bragg, April 9, 1974, Military History Institute, Carlisle Barracks, Creighton W. Abrams Papers.
18. *Ranger Handbook (SH 21-76)* found at http://www.africom.mil/WO-NCO/DownloadCenter/%5 C40Publications/Ranger%20Handbook.pdf accessed on December 31, 2011.
19. Hall, *The Ranger Book,* p. 4.
20. Ibid, p. 5.
21. Keith Nightingale kept a copy of the Ranger Creed that was posted in 1974 throughout the Battalion in Fort Benning and Fort Stewart. This insert was drawn from that copy.

CHAPTER 4

1. Interview with KC Leuer on August 10, 2011.
2. Since the establishment in 1951, Ranger School's focus has been on training the individual. The intent of the course is to train officers and noncommissioned officers who will return to their units to train their soldiers what they, themselves, had been taught. The school's goal is to graduate at least 3000 Ranges a year. 60% of the failures occur in the first three days with the swim, land navigation, physical fitness tests and foot marches accounting for most of the failures. 30% of all failures are due to personal reasons, and 10% are due to academic failures in the form of patrol, peer, and other observations. The Ranger Course is a nine week challenge to a man's self-confidence, physical fitness, and above all his determination to prevail. It was habitually held on the Harmony Church training. Today Harmony Church is no longer the home of the Ranger Training Brigade (RTB). The Armor Center has occupied the Harmony Church are since the Infantry and Armor Centers combined to make a Maneuver Center of Excellence at Fort Benning after 2005. The Ranger School is not organizationally affiliated with the operational Ranger units, instead it falls under the operational control of the Ranger Training Brigade which is subordinate to the US Army Training and Doctrine Command (TRADOC). Of note, generally an E4 Ranger who goes to Ranger School and graduates is automatically promoted to the Rank of Sergeant. Nowadays having to go to and pass the Sergeant, E-5 board is a requirement. Hall, *The Ranger Book,* p. 453 and Lock, John D., *The Coveted Black and Gold: A Daily Journal through the US Army Ranger School Experience,* Xlibris Corporation at www. Xlibris.com,2001, p. 30.
3. The M2 60mm lightweight mortar is a smoothbore, muzzle-loading, high angle of fire weapon used for close-in support of ground troops. It weighed 42 pounds replaced the M19 Mortar which weighed 52 pounds. Both of these weapon systems have been replaced by the M224 Mortar which ironically weighs 47 pounds in total for the system; five pounds more than the M2. It is likely that the Rangers of 1974 carried a mix of M2 and M19 Mortar Systems. The concept for development of the lightweight mortar was that it would bridge the gap between the hand grenade and the much heavier 81mm mortar system.

4. The recoilless rifle is a lightweight weapon that fires a heavier projectile that would be impractical to fire from a recoiling rifle. The key difference between rocket launcher and recoilless rifles is that the munitions fired from rifles have no propulsion of its own – once out of the rifle, it does not accelerate further, like a missile or rocket would once fired from its launcher. The first recoilless gun (a smooth bore variant of the rifle) was developed by the US Navy just prior to World War I. This gun, called the Davis Gun which was used experimentally against zeppelins and submarines. Today the Carl Gustav fires 84mm man-portable anti-tank- munitions.

5. The Ranger Support Element (RSE) is not organic (Riggers, Truck Drivers, Maintenance, etc.) to the Battalion. It is coordinated thru individual post memorandums of understanding (MOU) that provides the specific Battalion (1st, 2nd, or 3rd) with the necessary requirements to meet mission/training demands. It does not deploy with the Battalion to combat.

6. http://www.arp.sprnet.org/admin/FORMS/AT_RISK_CRITERIA.htm accessed on September 11, 2012.

7. Operational readiness training tests (ORTT) were used to evaluate units, teams and gunners on their ability to conduct operations under simulated combat conditions. The Army Training Test (ATT) is much like the ORTT for it was used against the tested units, most of the time battalion and brigades as a measured test not against an absolute standard but against a ladder of achievement set for in the Army Training Program. Memorandum for Record: Army Training Test Concept dated January 7, 1974, p. 1 found at http://usacac.army.mil/cac2/CSI/docs/Gorman/03_DCST_1973_77/04_74_ArmyTngTests_7Jan.pdf accessed on March 23, 2012.

8. Hall, *The Ranger Book,* p. 432.

9. Ranger Classes were numbered by the number in sequence for that year. Habitually there are 10-12 classes per year and Hawks was 6-75 meaning it began sometime in June 1975.

10. A C-130 Hercules has the capacity to jump 64 paratroopers on each aircraft. There are habitually 2 jumpmasters (one primary and one assistant) per aircraft and two safeties per aircraft who did not jump but assisted the jumpmasters in their duties.

11. Novice jumpers have 5 jumps and wear the plain silver airborne wings. Senior parachutists have at least 30 jumps logged to include 15 with combat equipment, two night jumps in which one has to be as a jumpmaster of a stick of paratroopers, and two mass tactical jumps in units consisting of a battalion or larger; a separate company, or an organic staff of a regimental size or larger. Finally to earn one's silver wings with a star atop denoting a senior parachutes rating you one had to serve for a total of 24 months in an airborne unit. To become Master parachutists rated one has to have at least 65 jumps, 25 of which required combat equipment, four night jumps with one as a jumpmaster, five mass tactical jumps and 36 months (nonconsecutive) with duty in an airborne unit. Once these criteria are met the paratrooper is authorized to wear wings with a star and a wreath encircling the star over the wings denoting his/her master parachutist status. Both senior and master rated parachutists have to be jumpmaster qualified. Soldiers who complete airborne jumps into combat zones can wear the appropriate level set of Parachutist wings with a bronze star for each jump.

12. For more on Rokosz and his command in Desert Shield and Desert Storm see Dominic J. Caraccilo, *The Ready Brigade of the 82nd Airborne in Desert Storm: A Combat Memoir of a Headquarters Company Commander,* Jefferson, North Carolina: McFarland & Company, 1993.

13. Hall, *The Ranger Book,* p. 439.

14. This attitude was appropriate for the era but would fall by the wayside years later with the introduction of the Multiple Integrated Laser Engagement System or MILES and other constructive type training aides. This system and others allows for a force on force exercise and requires large amount of blank ammunition.

15. Kirkham, R. R., Chamness M. A., Driver, C. J. and Barfuss, B. C., "Air Quality and Road Emission Results for Fort Stewart, Georgia," *Pacific Northwest National Laboratory* found at http://www.pnl.gov/main/publications/external/technical_reports/PNNL-14933.pdf (February 2005) assessed on June 18, 2014, p. 1.1.

16. Hall, *The Ranger Book,* p. 449.

17. The six colors were red, white, blue, green, orange and khaki. These are the colors on the modern day Ranger flash, the background patch that is sewn to the front of the tan beret that is used to place the rank of officers upon or the distinctive unit insignia (DUI) where an enlisted man dons his unit crest.

18. Secretary of the Army Howard H. Callaway, letter to General Creighton W. Abrams, Washington, DC, August 22, 1974, Military History Institute, Carlisle Barracks, Pennsylvania, Creighton W. Abrams Papers.

CHAPTER 5

1. http://www.armyranger.com/index.php/history/modern-era/vietnam accessed on September 14, 2012.
2. Exercise REFORGER (return of forces to Germany) was an annual exercise conducted during the Cold War by the North American Treaty Organization (NATO). The exercise was intended to ensure that NATO had the ability to quickly deploy forces to West Germany in the event of a conflict with the Warsaw Pact. The REFORGER exercise was first conceived in 1967. The Johnson administration announced plans to withdraw approximately two divisions from Europe in 1968. As a demonstration of its continuing commitment to the defense of NATO and to illustrate its capability of rapid reinforcement, a large scale force deployment was planned that would deploy a division or more to West Germany in a regular annual exercise. The first such exercise was conducted beginning on January 6, 1969. These exercises continued annually past the end of the Cold War, except for the year 1989, until 1993. REFORGER 1988 was billed as the largest European ground maneuver since the end of World War II as 125,000 troops were deployed. REFORGER was not merely a show of force—in the event of a conflict, it would be the actual plan to strengthen the NATO presence in Europe. In that instance, it would have been referred to as Operation REFORGER. Important components in REFORGER included the Military Airlift Command, the Military Sealift Command, and the Civil Reserve Air Fleet. The US Army also increased its rapid-reinforcement capability by prepositioning huge stocks of equipment and supplies in Europe at different forward located sites. The maintenance of this equipment provided extensive on-the-job training to reserve-component support units. Operation Bright Star, which still exists today, is the biannual deployment of American army and Air Force units to Egypt serves much the same purpose as REFORGER did. The 1975 REFORGER included the 3rd Infantry Division (Mechanized), 2nd Armored Cavalry Regiment, 3rd Armored Division, 1st Infantry Division (Mechanized), and the 1st Cavalry Division as well as the US Army Rangers. *The Stars and Stripes*, Vol. 47, No. 147, Sept. 12, 1988.
3. There are rigid regulations regarding the earning and wearing of foreign army jump wings. The jump has to be an interoperable operation with the foreign unit usually with a mix of US and host nation paratroopers jumping from the foreign Army's aircraft. There has to be a formal agreement with the two countries and the parameters of how many jumps that occur (can be as few as one) has to be agreed upon in writing. Most importantly, the foreign Army has to authorize the US parachutist to wear its nation's airborne wings and the awarding of such is usually done after the successful jump. The US soldier is only authorized to wear the foreign jump wings on his or hers dress uniform. More on the authorization of foreign wings can be found at https://www.hrc.army. mil/site/Active/TAGD/awards/Foreign_Badges_Appendix/Appendix_D.pdf.
4. Both of these safety features are common on present day parachutes. The safety wire ensure that the static line which when pulled to deploy the parachute upon exiting the aircraft doesn't come unhooked from the aircraft – if it did it the parachute wouldn't be pulled from its tray as the parachutist existed the aircraft causing him to free fall without a parachute. The anti-inversion prevents the parachute from inverting upon descent once it opens. If it did invert the parachutist wouldn't have any lift and fall freely to the ground.
5. The de Havilland Canada DHC-3 Otter is a single-engine, high-wing, propeller-driven, short take-off and landing (STOL) aircraft (a term used to describe aircraft with very short runway requirements) developed by de Havilland Canada. It was conceived to be capable of performing the same roles as the earlier and highly successful Beaver, but was overall a larger aircraft.
6. The Ferret armored car, also commonly called the Ferret Scout car, is a British armored fighting vehicle designed and built for reconnaissance purposes. The Ferret was produced between 1952 and 1971 by the UK Company Daimler. It was widely adopted by regiments in the British Army as well as Commonwealth countries throughout the period.

7. The PRC-74B is a solid state portable high frequency (HF) transceiver for special operations forces. It was built by Hughes and issued 1966.

8. A zone reconnaissance is conducted to obtain information on enemy. Zone reconnaissance techniques include the fan method. Using this method the leader first selects a series of rally points (ORP) throughout the zone. The platoon establishes security at the first ORP. Each reconnaissance and security (R&S) team moves from the ORP along a different fan-shaped route that overlaps with others to ensure reconnaissance of the entire area. The leader maintains a reserve at the ORP. When all R&S teams have returned to the ORP, the platoon collects and disseminates all information to every soldier before moving on to the next ORP. *Field Manual (FM) 7-8, Infantry Rifle Platoon and Squad* (Washington, DC: Department of the Army, April 22, 1992, pp. 3-12.

9. A Vulcan gun is a M163 Vulcan Air Defense System (VADS) which is a self-propelled anti-aircraft gun (SPAAG). The M168 gun is a variant of the General Dynamics 20 mm M61 Vulcan rotary cannon, the standard cannon in most US combat aircraft since the 1960. It is also mounted on either an armored vehicle or a trailer in the ground mode found at http://www.reference.com/browse/M163_VADS accessed April 21, 2012.

10. Black out drive is what soldiers use to drive in the dark. Tactical vehicles are equipped with low wattage marker, taillights, and a driving light that casts just a small amount of white light in front of the vehicle. These blackout lights work well with Night Vision Devices (NVDs).

11. A "Rat Rig" is a communications console that sat in the back of a truck.

12. The Commanding General of the 101st Airborne Division at that time was Major General John A. Wickham who went on to become a four star general and Chief of Staff of the Army.

13. http://specialforces78.com/about-the-special-forces/ranger/ accessed on September 12, 2012.

14. Sanctions imposed after Iranian students stormed the US embassy and took diplomats hostage in 1979 included a ban on most US-Iran trade. Goods or services from Iran were not allowed to be imported into the United States, directly or through third countries, with the following exceptions: gifts valued at $100 or less; information or informational materials; foodstuffs intended for human consumption; certain carpets and other textile floor coverings. "Factbox: US EU, and U.N. sanctions against Iran" Reuters web page found at http://www.reuters.com/article/2011/01/21/us-iran-nuclear-sanctions-idUSTRE70K03O20110121, January 20, 2011accessed on May 6, 2012.

15. Stephen Kinzer, "Inside Iran's Fury, Scholars Trace the Nation's Antagonism to its History of Domination by Foreign Powers," Smithsonian magazine, October 2008 found at http://www.smithsonianmag.com/people-places/iran-fury.html accessed on May 6, 2012.

16. John Skow, "The Long Ordeal of the Hostages," *Time,* January 26, 1981 found at http://www.time.com/time/magazine/article/0,9171,954605-3,00.html accessed on May 6, 2012.

17. The five airmen who died that night were Major Richard L. Bakke, Major Harold L Lewis Jr., Technical Sergeant Joel C. Mayo, Major Lyn D. McIntosh, and Captain Charles T. McMillan. The three US Marines casualties were Sergeant John D. Harvey, Corporal George N. Holmes Jr., and Staff Sergeant Dewey L Johnson. Richard A. Gabriel, A., *Military Incompetence: Why the American Military Doesn't Win*, New York, New York: Hill and Wang, 1985, pp. 106–116

18. http://www.enotes.com/topic/Operation_Eagle_Claw accessed on September 14, 2012.

19. Each of the EC130E were equipped with a pair of collapsible fuel bladders containing 6,000 US gallons of jet fuel, http://www.digplanet.com/wiki/Operation_Eagle_Claw accessed on September 15, 2012.

20. http://www.rtbot.net/Operation_Eagle_Claw accessed on September 15, 2012.

21. William A. Cohen, *The Art of the Strategist: 10 Essential Principles for Leading Your Com*pany to Victory, New York: AMACOM, 2004, p. 106.

22. Hall, *The Ranger Book,* p. 467.

23. http://www.armyranger.com/index.php/history/modern-era/iran accessed on September 15, 2012.

24. The word "haboob" comes from the Arabic word *habb*, meaning "wind." A haboob is a wall of dust as a result of a microburst or downburst. Air is forced downward and then pushed forward by the front of a thunderstorm dragging dust and debris with it, as it slowly travels across the terrain.

25. Which powers the number-one automatic flight control system and a portion of the primary flight controls, *Bluebeards 2* and *8*, which were later abandoned, now serve with the Iranian Navy.

26. Mark Bowden, *Guests of the Ayatollah: The First Battle In America's War with Militant Islam*, New York: Atlantic Monthly Press, 2006, p. 432.

27. http://8thattacksqdnassoc.tripod.com/AfterVietnam.html accessed on September 15, 2012.

28. Ibid.

29. The author was a first year plebe at the US Military Academy at West Point where the 44 hostages stayed during the repatriation process. Saluting the droves of busses transporting the returning Americans and then dining with them in the West Point mess hall was a memorable event.

30. James Bancroft. "The Hostage Rescue Attempt in Iran, 24–25 April 1980" found at http://rescue-attempt.tripod.com/id1.html. Accessed on May 6, 2012.

31. http://www.military.com/HomePage/UnitPageHistory/1,13506,100221%7C700109,00.html accessed on September 15, 2012.

32. http://wwwamericanpatriot-vance.blogspot.com/2009_01_01_archive.html accessed on September 15, 2012.

33. By a vote of 108 in favor to 9 (Antigua and Barbuda, Barbados, Dominican Republic, El Salvador, Israel, Jamaica, Saint Lucia, Saint Vincent and the Grenadines, and the United States) voting against, with 27 abstentions, the United Nations General Assembly adopted General Assembly Resolution 38/7 which "deeply deplores the armed intervention in Grenada, which constitutes a flagrant violation of international law and of the independence, sovereignty and territorial integrity of that State". The government of China termed the United States intervention an outright act of hegemony. The USSR government observed that Grenada had for a long time been the object of United States threats, that the invasion violated international law, and that no small nation not to the liking of the United States would find itself safe if the aggression against Grenada was not rebuffed. The governments of some countries stated that the United States intervention was a return to the era of barbarism. The governments of other countries said the United States by its invasion had violated several treaties and conventions to which it was a party. United Nations General Assembly resolution 38/7, page 19," United Nations. November 2, 1983 found at http://www.un.org/depts/dhl/resguide/r38.htm, retrieved on May 11, 2012 and *United Nations Yearbook*, Volume 37, 1983, Department of Public Information, United Nations, New York.

34. http://www.suasponte.com/m_grenada.htm accessed on September 15, 2012.

35. http://www.shadowspear.com/special-operations/1468-operation-urgent-fury.html accessed on September 15, 2012.

36. *Howard Zinn, A People's History of the United States: 1492-Present, New York:* Harper Perennial, 2005, Extract found at http://libcom.org/history/1983-the-us-invasion-of-grenada accessed on May 10, 2011.

37. Ronald H. Spector, US Marines in Grenada 1983, Washington DC: History and Museums Division, 1987, p. 3.

38. Most combat jumps into enemy territory have the parachutists already donned in their parachutes before loading. The preferred method is always to airland for it was less risky and this is what the Rangers had planned at Point Salines. The option to jump via parachute was always there and the Rangers were ready for it in this mission. Therefore when the word came down that the airstrip was blocked the decision to jump was made and the Rangers donned their parachute. Lock, *Rangers in Combat,* p. 208.

39. Lock, *Rangers in Combat,* pp. 209-211.

40. http://www.armyranger.com/index.php/history/modern-era/grenada accessed on September 15, 2012.

41. Lock, *Rangers in Combat,* pp. 212.

42. http://www.shadowspear.com/special-operations/1468-operation-urgent-fury.html accessed on September 15, 2012.

43. Kent T Woods, "Rangers Lead the Way: The Vision of General Creighton W. Abrams, US Army War College Strategy Research Project, Carlisle Barracks, Pennsylvania, July, 14, 2003, p. 9.

44. Lock, *To Fight with Intrepidity,* p. 440.

45. http://www.military.com/HomePage/UnitPageHistory/1,13506,100221%7C700109,00.html accessed on September 15, 2012.

46. http://www.specialoperations.com/Army/Rangers/History.htm accessed on September 15, 2012.

47. Kent T Woods, "Rangers Lead the Way: The Vision of General Creighton W. Abrams, US Army War College Strategy Research Project, Carlisle Barracks, Pennsylvania, July, 14, 2003, pp. 13-14.

48. Dominic J. Caraccilo, *Beyond Guns and Steel: A War Termination Strategy,* Santa Barbara, California: Praeger Security International, 2011, p. 84.

49. http://www.scribd.com/doc/31125838/Operation-JustCause, p. 15 accessed on September 15, 2012.

50. Ronald Cole, "Operation Just Cause: Panama," Joint History Office, Office of the Chairman of the Joint Chiefs of Staff, Washington, DC, 1995, p. 7.

51. http://www.scribd.com/doc/31125838/Operation-JustCause, p. 16 accessed on September 15, 2012.

52. Ibid, pp. 18-20.

53. Cole, p. 35.

54. As part of the Joint Special Operations Command (JSOC) each element or task force is given a color coded name to represent them in operations. The Rangers are "Red," Delta Force is "Green," Seal Team Six is "Blue," the 160th Special Operation Aviation Regiment is "Brown," and the Air Force Special Operations element is "Silver." There are other Special Mission Units (SMU) also associated with JSOC and they remain classified.

55. http://www.armyranger.com/index.php?option=com_content&view=article&id=40&Itemid=60 accessed on September 15, 2012.

56. Lock, *To Fight with Intrepidity,* pp. 472-473.

57. Eglin is the home of the Air Force special operations unit.

58. http://www.suasponte.com/modernrang.htm accessed on September 15, 2012.

59. http://www.johndlock.com/tfwi-13.html accessed on September 15, 2012.

60. Douglas Jehl, "Ranger Force Bore Brunt of Panama Toll," Los Angeles Times, January 7, 1990 found at http://articles.latimes.com/1990-01-07/news/mn-445_1_rio-hato/2 accessed on May 31, 2012.

61. http://www.armyranger.com/index.php?option=com_content&view=article&id=40&Itemid=60 accessed on September 16, 2012.

62. Ibid.

63. Douglas Jehl, "Ranger Force Bore Brunt of Panama Toll," Los Angeles Times, January 7, 1990 found at http://articles.latimes.com/1990-01-07/news/mn-445_1_rio-hato/2 accessed on May 31, 2012.

64. The author wrote his first book on this very topic titled *The Ready Brigade of the 82nd Airborne in Desert Storm: A Combat Memoir of a Headquarters Company Commander,* Jefferson, North Carolina: McFarland & Company, 1993.

65. Hall, *The Ranger Book,* p. 494.

66. Woods, p. 14.

67. *Improving the Prospects for Future International Peace Operations*, Washington, DC: Office of Technology Assessment, 1995, p. 44.

68. Forrest E. Morgan, Karl P. Mueller, Evan S. Medeiros, Kevin L. Pollpeter and Roger Cliff, *"Dangerous thresholds: managing escalations in the 21st century,"* Santa Monica, California: *Rand Corporation Project Air Force,* 2008, pp. 215-216.

69. Senate Report and Bill Gertz, "Aspin's decision on tanks was political," *The Washington Times* (October 3, 1995) found at http://www.netnomad.com/powell.html accessed on May 3, 2012.

70. Morgan, et. al, *Dangerous thresholds,* pp. 207-208.

71. http://www.warandtactics.com/smf/war-conflicts-20th-century-discussions/battle-of-moga-dishu-october-2-3-1993-(article)/ accessed on September 16, 2012.

72. https://www.armyranger.com/ranger/index.php/history/modern-era/somalia access on June 26, 2014.

73. http://www.leatherneck.com/forums/archive/index.php/t-17200.html accessed on September 15, 2012.

74. Rangers often referred to Delta Operators as "The Hardy Boys."

75. Morgan, et. al, *Dangerous thresholds,* p. 216.

76. http://www.leatherneck.com/forums/archive/index.php/t-17200.html accessed on September 15, 2012.

77. For an excellent overview of what occurred in Mogadishu these two fateful days in October 1993 see Rick Atkinson, "Firefight in Mogadishu: The Last Mission of Task Force Ranger," *The Washington Post,* January 31, 1994.

78. http://www.armyranger.com/index.php/history/modern-era/somalia accessed on September 16, 2012.

79. Lock, *To Fight with Intrepidity* found at http://www.johndlock.com/#!tfwi-extract---tf-ranger-somalia/crl8 accessed on June 26, 2014.

80. Robin Moore and Michael Lennon, *The Wars of the Green Berets: Amazing Stories from Vietnam to the Present*, New York, New York: Skyhorse Publishing, 2007, p. 28.

81. Swenson, Derek, In the Eyes of the Beholder, Xlibris found at http://books.google.com/books?id=8WJNAAAAQBAJ&pg=PT220&lpg=PT220&dq=helicopters+lifted+off+at+1532.&source=bl&ots=Hs76GynPuv&sig=Jztn9zBOE8QCJyVb5-w1rtBTyso&hl=en&sa=X&ei=RV6sU9iQH9CoyATL3oCYDA&ved=0CEYQ6AEwBA#v=onepage&q=helicopters%20lifted%20off%20at%201532.&f=false (2010).

82. Rick Atkinson, "Firefight in Mogadishu: The Last Mission of Task Force Ranger," *The Washington Post,* January 31, 1994, p. A01.

83. http://www.armyranger.com/index.php/history/modern-era/somalia accessed on September 16, 2012.

84. http://www.suasponte.com/m_somalia.htm accessed on September 16, 2012.

85. Rick Atkinson, "Firefight in Mogadishu: The Last Mission of Task Force Ranger," *The Washington Post,* January 31, 1994, p. A01.

86. Mark Bowden, "Black Hawk Down: An American War Story," *The Philadelphia Enquirer,* November 23, 1997, found at http://inquirer.philly.com/packages/somalia/nov23/default23.asp accessed on June 2, 2012.

87. https://www.armyranger.com/ranger/index.php/history/modern-era/somalia accessed on June 26, 2014.

88. Lock, *To Fight with Intrepidity* found at http://www.johndlock.com/#!tfwi-extract---tf-ranger-somalia/crl8 accessed June 26, 2014.

89. http://www.sfa-72.com/html/moh_gary_gordon-_hon_life_member_sfa-72.html accessed on June 27, 2014.

90. http://www.armyranger.com/index.php/history/modern-era/somalia accessed on June 27, 2014.

91. Mark Bowden, "A second crash site, and no escape," *Black Hawk Down: An American War Story* excerpt in the *Philadelphia Inquirer* found http://inquirer.philly.com/packages/somalia/nov23/default23.asp, November 23, 1997 accessed on June 27, 2014.

92. Ibid.

93. http://www.johndlock.com/#!tfwi-extract---tf-ranger-somalia/crl8 accessed on June 27, 2014.

94. While much has been written on the heroics displayed by the Delta Operatives and 160 pilots during the Somalian fight, there were a set of 75th Ranger heroes that made the ultimate sacrifice. They included from the 3rd Battalion, 75th Ranger Casualties included: Corporal James M. Cavaco, Sergeant James C. Joyce, Specialist Richard W. Kowalewski, Sergeant Dominick M. Pilla, Sergeant Lorenzo M. Ruiz and Corporal James E. Smith found at http://www.suasponte.com/m_fallen.htm accessed on September 15, 2012.

95. The author's brother, Maj. (retired) Ed Caraccilo was a Specialist in the 1st Ranger Battalion at the time of the potential Haiti invasion and was stationed on the Aircraft Carrier readying to fast rope into the Presidential Plaza.

96. Hall, *The Ranger Book,* pp. 488-489.

CHAPTER 6

1. William Mcraven, *Spec Ops: Case Studies in Special Operations Warfare: Theory and Practice*, New York, New York: Presidio Press, 2006, p. 8-23.

2. Mir Bahmanyar, *Shadow Warriors: A History of the US Army Rangers*, New York, New York: Random House Inc., 2005. p. 180.

3. Linda D. Kozary, "Defense Leaders Uphold Army's Black Beret Decision," *American Forces Press Service,* March 16, 2001 found at http://www.defense.gov/news/newsarticle.aspx?id=45776 accessed on June 29, 2104.

4. http://www.socnet.com/showthread.php?t=2736 dated March 17, 2001 accessed on July 10, 2014..

5. Of the thousands killed, 55 were active duty military Andrea Stone, "Military's aid and comfort ease 9/11 survivors' burden," *USA Today*, August 20, 2002.
6. David Ferguson, "Bush Says Slow Reaction on 9/11 was Deliberate," *The Raw Story,* June 29, 2011 found at http://www.rawstory.com/rs/2011/07/29/bush-says-slow-reaction-on-911-was-deliberate-decision, accessed on June 9, 2012 and Amy Woods, "Rumsfeld: Bush Response to 9/11'Measured,' 'Decisive," *Newsmax* September 11, 2001, found at http://www.newsmax.com/InsideCover/Sept11-Rumsfeld/2011/09/11/id/410475 accessed on September 15, 2012.
7. Kathleen T. Rhem, "Bush: No distinction between attackers and those who harbor them," American Forces Press Service, Washington DC, September 11, 2001 found at http://www.defense.gov/news/newsarticle.aspx?id=44910, accessed on June 9, 2012.
8. The following is a list of plots and attacks by Al-Qaeda leading to the 9/11 Attack, "Al-Qaida timeline: Plots and Attacks" found at http://www.msnbc.msn.com/id/4677978/ns/world_news-hunt_for_al_qaida/t/al-qaida-timeline-plots-attacks/, accessed on June 3, 2012:
— December 29, 1992
In the first al-Qaida attack against US forces, operatives bomb a hotel where US troops -- on their way to a humanitarian mission in Somalia -- had been staying. Two Austrian tourists are killed. Almost simultaneously, another group of al-Qaida operatives are caught at Aden airport, Yemen, as they prepare to launch rockets at US military planes. US troops quickly leave Aden.
— February 26, 1993
The first World Trade Center attack and the first terrorist attack on the US -a bomb built in nearby Jersey City is driven into an underground garage at the trade center and detonated, killing six and wounding 1,500. Ramzi Yousef, nephew of Khalid Sheik Mohammed, master-minds the attack, working with nearly a dozen local Muslims. While US officials disagree on whether Osama bin Laden instituted the attack and Yousef denies he has met bin Laden, the CIA later learns that Yousef stayed in a bin Laden-owned guest house in Pakistan both before and after the attacks.
— April – June 23, 1993
Militants plan a series of near simultaneous bombings in New York-among the targets were promi-nent New York monuments: the Lincoln and Holland tunnels; the George Washington Bridge; the Statue of Liberty; the United Nations; the Federal Building at 26 Federal Plaza; and finally, one in the Diamond District along 47th Street. On June 23, as terrorists mix chemicals for the bombs and FBI agents raid their warehouse and arrest twelve.
— May - July 28, 1993
After two months of planning, Ramzi Yousef, mastermind of the World Trade Center bombing, travels to Karachi, the hometown of Benazir Bhutto-then former prime minister of Pakistan, who is seeking to regain her old job. He and two others are in the process of planting a remote control bomb on the road when the aging Soviet detonator obtained in Afghanistan explodes in his face, ending the plot. Financing for the bombing comes from radical Islamic groups in Pakistan, according to Bhutto.
— June 1993
Al-Qaeda reportedly attempts to assassinate then Jordanian Crown Prince Abdullah.
— October 3-4, 1993
In a battle for the streets of Mogadishu, Somalia, a unit of US special operations forces gets pinned down after two US helicopters are shot out of the sky. Eighteen Americans die, killed by Somalis reportedly trained by al-Qaida. The attack leads to the US withdrawal from Somalia, a move hailed by bin Laden as a great victory for the Islamic world.
— March 11, 1994
Led by Ramzi Yousef, a group of Islamic militants hijack a delivery truck in downtown Bangkok, strangle the driver and load a one-ton bomb on board. Their target: the Israeli embassy. However the truck has an accident and the hijacker abandons it, leading to the discovery of the bomb and driver's body. Several of the plotters are arrested, but Yousef escapes again.
— June 1994
Imad Mugniyeh, the military chief of Hezbollah during its 1980's attacks on US personnel, meets secretly with Bin Laden in Khartoum. At that point, Mugniyeh is the most wanted terrorist in the world for his role in the Beirut embassy and Marine Barracks bombing and begins advising Bin Laden on planning. Ali Mohamed, the al-Qaida security director at the time, later tells US offi-cials that Mugniyeh told bin Laden how the Marine bombing in Beirut led to the US withdrawal

from Lebanon and how such a campaign could eventually lead to a similar route of US troops in Saudi Arabia and the whole Islamic world.
— June 20, 1994
Ramzi Yousef, working with the People's Mujahedin of Iran, blows up the Shrine of Reza in Mashad, Iran. The explosion took out the entire wall of the mausoleum, killing 26 pilgrims, mostly women. At the time, Yousef was motivated as much by hatred of Shiite Muslims as by hatred of America.
— November 12-14, 1994
Extremists working for bin Laden conduct extensive surveillance of President Bill Clinton and his party during a state visit to Manila in anticipation of mounting an assassination attempt when Clinton returns to the Philippine capital in November 1996 for an already scheduled APEC summit. Bin Laden orders al-Qaida to use still and video cameras to follow Clinton and Secret Service personnel. The Secret Service later learns from an al-Qaida defector that the surveillance was extensive, and the tapes along with maps and notes were sent to bin Laden, who was then living in Sudan. The Secret Service was unaware of the surveillance although there was some concern at the time that the president was exposed during the trip.
— December 8, 1994 - Jan. 5, 1995
Ramzi Yousef rents an apartment in the Dona Josefa apartment complex on Quirino Boulevard, in Manila, Philippines, believing that Pope John Paul II will take that route on his way to a huge outdoor mass planned for Jan. 15. The apartment is only 500 feet from the Manila home of the Vatican ambassador to the Philippines where the Pope will stay during his 5-day visit to the country. In addition, he rents a beach house to train his compatriots for the attack and purchases two Bibles, a crucifix, a large poster of the Pope, several priests' garments — accurate down to the tunic buttons and confessional manuals. The plan, investigators said, was to place a bomb under a manhole cover along Quirino Boulevard. The attack is thwarted when bomb-making materials catch fire in the sink of the apartment kitchen. As it turns out, the pope travels to the Mass by helicopter.
— December 10, 1994
As part of the planning for the Day of Hate Yousef plants a crude bomb on board a Philippines Airlines plane from Cebu City, the Philippines, to Tokyo. When the bomb detonates, it kills a Japanese businessman, and forces the 747 to land in Okinawa.
— January. 21-22, 1995
In what would have been an attack with a higher death toll than the Sept. 11 attacks, bombs placed on board 11 jumbo jets are to be detonated by timing devices as the planes fly over the Pacific, killing an estimated 4,000 people. Most of the jets are to be American carriers and most of the dead would have been Americans. The bombs would have been timed to go off over a number of hours to heighten the terror. The plan, called the Day of Hate, was conceived by Ramzi Yousef, the mastermind of the first World Trade Center bombing and his uncle, Khalid Sheik Mohammed. Only a fire in Yousef's Manila apartment on Jan. 6 thwarts it. Mohammed later modifies the plan and takes it to Osama bin Laden. That modified plan becomes the blueprint for the Sept. 11, 2001 attacks.
— June 26, 1995
Less than an hour after Egypt's President Hosni Mubarak arrives in Addis Ababa to attend the Organization of African Unity summit, several members of the Egyptian Islamic Jihad, a group working with al-Qaida, attack his motorcade. Ethiopian forces kill five of the attackers and capture three others. Ethiopia and Egypt charge the government of Sudan, where bin Laden is living, with complicity in the attack and harboring suspects. Privately, Egyptian officials tell US intelligence they believe Bin Laden is behind the attack. Later, Egyptian officials learn that the terrorists had conducted surveillance of the last trip Mubarak had made to Ethiopia, just as they had with President Clinton.
— November 13, 1995
A truck bomb explodes outside the Saudi National Guard Communications Center in central Riyadh, killing five American servicemen and two Indian police. Four Saudi men, all self-described disciples of bin Laden, are quickly executed before the FBI can determine their ties to al-Qaida.
— June 25, 1996
In an attack whose authorship is still debated by intelligence and law enforcement officials, a truck bomb is detonated at the Khobar Towers complex in Saudi Arabia, killing 19 American servicemen and wounding 400. Although an indictment in early 2001 pins blame on Shiite Muslims backed

by Iran, many US officials still believe bin Laden is responsible. Bin Laden himself states in a 1997 interview, "Only Americans were killed in the explosions. No Saudi suffered any injury. When I got the news about these blasts, I was very happy."
— August. 8, 1998
Al-Qaeda sends suicide bombers into the US embassies in Nairobi, Kenya, and Dar es Salaam, Tanzania. Truck bombs kill more than 240 people, including 12 Americans at the Nairobi embassy. The attack results in the quick arrest of several of the bombers, but not the mastermind, Fazul Abdullah Mohammed. Also known as "Harun," Mohammed is involved in later al-Qaida attacks.
— January 1-3, 2000
US and Jordanian authorities thwart attacks planned to coincide with the Millennium celebrations. In mid-December, Jordanian authorities arrest more than 20 al-Qaida operatives who are planning to bomb three locations where American tourists gather: Mt. Nebo, where Moses first saw the Promised Land; the Ramada Hotel in Amman, a stopover for tour groups; and the spot on the Jordan River where tradition holds John the Baptist baptized Christ. Later in the month, US authorities seize Ahmed Ressam at a border crossing in Port Angeles, WA. He is carrying bomb-making equipment and later discusses his plan to blow up Los Angeles International Airport on New Year's Eve.
— January 13, 2000
The cross Africa Dakar-to-Cairo auto race is diverted after the US intelligence community receives word of a planned ambush in the African nation of Niger. Word of the planned ambush was passed to race organizers over the weekend shortly after it was received, leading to a suspension of the race and a massive airlift on Thursday. Cargo planes were flying some 1,365 crew members and 336 vehicles as well as tons of equipment from Niamey, capital of Niger, to Sabha in southern Libya.
— October 12, 2000
A bomb on board a small Zodiac-like boat detonates near the USS *Cole* in the port of Aden in Yemen, killing 17 US sailors and wounding scores more. The bombing also kills two al-Qaida operatives in the boat. The United States later learns the Cole was the second destroyer targeted by al-Qaida. The attack was originally planned for Jan. 3, 2000, when the USS The Sullivans was in the same port.
— September 9, 2001
Two Moroccan men, posing as television journalists, kill themselves and Ahmad Shah Massoud, leader of the Northern Alliance, at the alliance headquarters in the Panjshir Valley of Afghanistan. The killing of Massoud may have been the first part of the Sept. 11 attacks.
— September 11, 2001
Three hijacked planes are flown into major US landmarks, destroying New York's World Trade Center towers and plowing into the Pentagon. A fourth hijacked plane crashes in rural Pennsylvania, its target believed to have been the US Capitol. The death toll is nearly 10 times greater than any other terrorist attack in history and makes bin Laden, for the first time, a household name in the United States and the west.

9.   http://www.bharat-rakshak.com/SRR/Volume11/narayanan.html accessed on September 17, 2012.
10. J.T. Caruso, Acting Assistant Director, Counterterrorism Division, FBI, Federal Bureau of Investigation, Testimony before the Subcommittee on International Operations and Terrorism, Committee on Foreign Relations, Washington, DC: United States Senate, December 8, 2001 found at http://www.fbi.gov/news/testimony/al-qaeda-international accessed on June 29, 2014.
11. Lawrence J. Bevy, "Al-Qaeda: an organization to be reckoned with," Hauppauge New York: Nova Science Pub Inc., 2006.p. 56.
12. A Fatwa is typically a declaration by a recognized authority figure and while this declaration is not a binding or legal directive it is usually taken serious by those that respect the authority figure. "Al-Qaeda's Fatwa," PBS Newshour, February 23, 1998, found at http://www.pbs.org/newshour/terrorism/international/fatwa_1998.html, accessed on June 16, 2012.
13. Tommy Franks, *American Soldier*, New York, New York: Harper Collins, 2004, p.362 and Bob Woodward, *Plan of Attack*, New York, New York: Simon & Schuster, 2004, p.292.
14. Eric Schmitt, "Threats and responses: military spending: Pentagon contradicts general on Iraq occupation force's size," New York Times found at http://www.nytimes.com/2003/02/28/us/threats-responses-military-spending-pentagon-contradicts-general-iraq-occupation.html, February 23, 2003 accessed on June 30, 2014.

15. "Bush announces strikes against Taliban," The Washington Post, October 7, 2001 found at http://www.washingtonpost.com/wp-srv/nation/specials/attacked/transcripts/bushaddress_100801.htm accessed on June 30, 2014.

16. The AC-130H "Spectre" can be armed with two 20 mm M61 Vulcan cannons, one Bofors 40mm auto cannon, and one 105 mm M102 cannon. The upgraded AC-130U "Spooky" has a single 25 mm GAU-12 Equalizer in place of the Spectre's twin 20 mm cannons, as well as an improved fire control system and increased capacity for ammunition. Power is provided by four Allison T56-A-15 turboprops (standard for a C-130 Hercules). New AC-130J gunships based on MC-130J Combat Shadow II special operations tankers are planned. Found at http://www.funker530.com/ac-130-gunship/, accessed on June 30, 2014.

17. Retired Brig. Gen. Eric Smith, the former commander of the 187th Infantry Regiment, the same command (also known as the 3rd Brigade of the 101st Airborne Division) that the author commanded some years later in Iraq, termed the categories of a Soldier as killer, filler and fodder which puts requirements for a combat force clearly in perspective.

18. The Lockheed P-3 Orion is a four-engine turboprop anti-submarine and maritime surveillance aircraft developed for the United States Navy and introduced in the 1960s. Lockheed based it on the L-188 Electra commercial airliner. The aircraft is easily recognizable by its distinctive tail stinger or "MAD Boom", used for the magnetic detection of submarines. Over the years, the aircraft has seen numerous design advancements, most notably to its electronics packages. The P-3 Orion is still in use by numerous navies and air forces around the world, primarily for maritime patrol, reconnaissance, anti-surface warfare and anti-submarine warfare. A total of 734 P-3s have been built, and by 2012, it will join the handful of military aircraft such as the Boeing B-52 Stratofortress which have served 50 years of continuous use with its original primary customer, in this case, the United States Navy. The US Navy's remaining P-3C aircraft will eventually be replaced by the Boeing P-8A Poseidon. Found at http://en.wikipedia.org/wiki/Lockheed_P-3_Orion, accessed on June 12, 2012.

19. Video of the Ranger jump onto Objective Rhino on October 19, 2001 can be found at http://www.youtube.com/watch?v=fcF2ctMdMsE.

20. Charles H. Briscoe, Richard L. Kiper, James A. Schroder and Kalev I. Sepp, "Weapon of Choice U.S. Army Special Operations Forces in Afghanistan," Leavenworth, Kansas: Combat Studies Institute Press, 2001, p. 98.

21. Sean M. Maloney, "Enduring the freedom: a rogue historian in Afghanistan, "Dulles, Virginia: Potomac Books, Inc., 2006, pp. 48-49.

22. Briscoe, et. al., p. 111.

23. http://en.inforapid.org/index.php?search=Operation%20Rhino accessed on July 1, 2014.

24. Ibid accessed July 1, 2014.

25. The author was a battalion commander for the 2nd Battalion, 503rd Infantry (Airborne) during the initial assault into Iraq where he commanded the force in the city of Kirkuk for a year in follow-on operations.

26. John D. Gresham, "The Haditha Dam Seizure – The Target," defensemedianetwork, May 1, 2010 found at http://www.defensemedianetwork.com/stories/hold-until-relieved-the-haditha-dam-seizure/, accessed on June 9, 2012.

27. Jeremy Bender, "ISIS is closing in on Iraq's most important dam," Business Insider found, June 25, 2014 at http://www.businessinsider.com/isis-is-still-threatening-iraqs-water-supply-2014-6 accessed on July 2, 2014.

28. Gresham, "The Haditha Dam Seizure – The Target," accessed on July 2, 2014.

29. Ibid.

30. Leigh Neville, Shawn Carpenter, Jim Wonacott, Jim Roots, Robby Carpenter, Road to Baghdad: Iraq 2003, Oxford, UK: Osprey Publishing, 2011.

31. Oliver North, American Heroes in Special Operations, Jarrell, Texas: Fidelis 2010, p. 61.

32. www.suasponte.com accessed on June 9, 2012.

33. Gresham, "The Haditha Dam Seizure – The Target," accessed on July 2, 201.

34. Ibid.

35. Ron Paul, Forums: Liberty Forest, April 2003 found at http://www.ronpaulforums.com/showthread.php?454471-Haditha-Dam-Part-II accessed on July 3, 2014

36. http://www.suasponte.com/m_iraq.htm accessed on September 17, 2012.
37. Gresham, "The Haditha Dam Seizure – The Target," accessed on July 2, 201.
38. Sean Naylor, *Not a Good Day to Die; The Untold Story of Operation Anaconda*, New York, New York: Berkley/Penguin Publishing, 2005, p. 90.
39. Mir Bahmanyar, *Shadow Warriors: A History of the US Army Rangers*, New York, New York: Random House Inc., 2005, p. 258.
40. Mir Bahmanyar, *US Army Ranger 1983-2002: Sua Sponte - Of Their Own Accord*, Oxford, UK: Osprey Publishing, 2003, p. 8.

CHAPTER 7

1. Lock, *Rangers in Combat,* p. 365.
2. The Headquarters and Headquarters Company (HHC) includes the Battalion Headquarters and the Staff, Medical, Maintenance, and Rigger Sections for the Battalion.
3. The RASP programs are an evolution from the RIP and ROP programs established to assess enlisted and officers explained earlier in this work. http://www.goarmy.com/ranger/heritage/regimental-special-troops-battalion.html accessed on September 17, 2012
4. http://www.reocities.com/collegepark/square/4420/A_Comprehensive_Look_at_Ranger_Training.htm accessed on September 17, 2012.
5. A vision can be political, religious, environmental, social, or technological in nature. By extension, a visionary can also be a person with a clear, distinctive, and specific outlook of the future, usually connected with advances in technology, organizational structure or social/political arrangements.
6. Stephen Peter Rosen, "The Future of War and the American Military," *Harvard Magazine,* May-June 2002, found at http://harvardmagazine.com/2002/05/the-future-of-war-and-th.html, accessed June 17, 2012.
7. Ibid.
8. Betsy Mason, "The Science of the Future of War," *Wired Science,* November 28, 2008 found at http://www.wired.com/wiredscience/2008/11/sex-and-war-exc/, accessed on June 17, 2012.
9. For analysis purposes it is instructive to look at the typical 21st Century Ranger today before delving into defining requirements for what's next.
The typical US Army Ranger at the time of this writing is as follows:
Average age: 24
Average height/weight: 5ft 9in/174lb
Time in service/time on station: 4½ years/ 2 years
Military training: Basic Combat Training and Advanced Individual Training, Airborne School (3 weeks), Ranger Indoctrination Program (1 month), Pre-Ranger Course (3 weeks), Ranger School (2 months), Ranger First Responder Medical Training (1 week), Primary Leadership Development Course (4 weeks) Experience: OEF/OIF deployments × 1–4, Joint Readiness Training Center rotation × 1, Joint Readiness Exercise × 1, live fire exercises × 25 Rank: About half are specialists (pay grade E-4)
Army Physical Fitness Test Score: 275 out of 300
Awards: Expert Infantryman Badge, Combat Infantryman Badge, Army Commendation Medal, Army Achievement Medal, GWOT Expeditionary and Service medals, Ranger Tab, Parachutist Badge
Other Statistics: Fewer than half are married; average number of children is 1.75; most have some college education; about half are Ranger qualified (earned Ranger Tab). Mir Bahmanyar, *Shadow Warriors: A History of the US Army Rangers*, New York, New York: Random House Inc., 2005, p. 333.
10. Kent T. Woods, "Rangers Lead the Way: The Vision of General Creighton W. Abrams: US Army War College Strategy Research Project, Carlisle Barracks, Pennsylvania, July, 14, 2003, p. 11.
11. Lawson G. Magruder, Lieutenant General (Retired), telephone interview by Kent T. Woods, March 5, 2003, p. 11.
12. Ibid, p. 12.

APPENDIX A

1 The US Army Ranger Association found at http://www.ranger.org/ accessed on June 23, 2012 and July 4, 2014. Used by permission of Gerard B. Nery, Jr.

APPENDIX B

1 Multiple publications list the "rules" in various forms. This list was taken from *US Special Operations Forces,* US International Business Publications, 2011, pp. 123-125 and from Lock, *The Coveted Black and Gold,* pp. 220-225.
2 Multiple venues list the "Standing Orders in various forms. This particular list was extracted from www.psywarriors.com/ranger.html accessed on February 2, 2012.

Bibliography

ARTICLES, BOOKS, MAGAZINES AND UNPUBLISHED DOCUMENTS

Aboul-Enein, Youssef, *Infantry Magazine*, September-October 2010 found at http://findarticles.com/p/articles/mi_m0IAV/is_3_99/ai_n56541299.

Abrams, Creighton W., General. Letter to Lieutenant General Samuel T. Williams, USA, Retired, Carlisle Barracks, Pennsylvania: History Institute, Creighton W. Abrams Papers, September 26, 1972.

Adams, John A. and Dethloff, Henry C., *Texas Aggies go to War: In Service of Their Country, College Station*, Texas: Texas A & M University Press, 2008.

Adkin, Mark, *Urgent Fury, The Battle for Grenada,* Lexington, Virginia: Lexington Books, 1989.

Aitken, Jonathan, *Nixon: A Life*, Washington DC: Public Affairs, Regency Publishing, Inc., 2008.

"Al Qaeda's Fatwah," *PBS Newshour*, February 23, 1998, found at http://www.pbs.org/newshour/terrorism/international/fatwa_1998.html, accessed on June 16, 2012.

Altieri, James, *The Spearheaders: A Personal History of Darby's Rangers,* Washington: DC: Zenger Publishing, Inc., 1960.

Ambrose, Stephen E., *D-Day: June 6, 1944, The Climatic Battle of World War II,* New York: Simon & Schuster, 1995.

American Archive Documents of the American Revolution, 1774-1776, produced by Northern Illinois University Library p. V2:1847 found at http://lincoln.lib.niu.edu/cgi-bin/amarch/getdoc.pl?/var/lib/philogic/databases/amarch/.5774 accessed on February 2, 2012.

Atkinson, Rick, "Firefight in Mogadishu: The Last Mission of Task Force Ranger," *The Washington Post,* January 31, 1994.

Bahmanyar, Mir, Shadow *Warriors: A History of US Army Rangers*, New York, New York: Osprey Publishing, 2005.

Bahmanyar, Mir, *US Army Ranger 1983-2002: Sua Sponte – Of Their Own Accord*, Oxford, UK: Osprey Publishing, 2003.

Bancroft, James, "The Hostage Rescue Attempt in Iran, 24–25 April 1980" found at http://rescueattempt.tripod.com/id1.html accessed on May 6, 2012.

Barbour, Philip L., ed., *The Complete Works of Captain John Smith,* Chapel Hill, North Carolina: University of North Carolina Press, 1986.

Bender, Jeremy, "ISIS is closing in on Iraq's most important dam," *Business Insider* found, June 25, 2014 at http://www.businessinsider.com/isis-is-still-threatening-iraqs-water-supply-2014-6 accessed on July 2, 2014.

Bevy, Lawrence J. "Al-Qaeda: an organization to be reckoned with," Hauppauge New York: Nova Science Pub Inc., 2006.

Black, Robert W., *Ranger Dawn: the American Ranger from the Colonial Era to the Mexican War*, Mechanicsburg, Pennsylvania: Stackpole Books, 2009.

Black, Robert, *Rangers in Korea*, New York, New York: Ivy Books, Random House, 1989.

Black, Robert, *Rangers in World War II*, New York, New York: Ballentine Books, Random House, 1992.

Boatner, John M., *Military Customs and Traditions*, Westport Connecticut: Greenwood Press, 1976.

Bowden, Mark, "Black Hawk Down: An American War Story," *The Philadelphia Enquirer*, November 23, 1997, found at http://inquirer.philly.com/packages/somalia/nov23/default23.asp accessed on June 2, 2012.

Bowden, Mark, *Guests of the Ayatollah: The First Battle In America's War with Militant Islam*, New York: Atlantic Monthly Press, 2006.

Breuer, William H., *The Great Raid on Cabanatuan*, New York, New York: John Wiley & Sons, 1994.

Briscoe, Charles H., Kiper, Richard L., Schroder, James A. and Sepp, Kalev I., "Weapon of Choice U.S. Army Special Operations Forces in Afghanistan," Leavenworth, Kansas: Combat Studies Institute Press, 2001.

"Bush announces strikes against Taliban," The Washington Post, October 7, 2001 found at http://www.washingtonpost.com/wp-srv/nation/specials/attacked/transcripts/bushad-dress_100801.htm accessed on June 30, 2014.

Caraccilo, Dominic J., *Beyond Guns and Steel: A War Termination Strategy*, Santa Barbara, California: Praeger Security International, 2011.

Caraccilo, Dominic J., *The Ready Brigade of the 82nd Airborne in Desert Storm: A Combat Memoir of a Headquarters Company Commander*, Jefferson, North Carolina: McFarland & Company, 1993.

Caruso, J.T., Acting Assistant Director, Counterterrorism Division, FBI, Federal Bureau of Investigation, *Testimony Before the Subcommittee on International Operations and Terrorism, Committee on Foreign Relations*, Washington, DC: United States Senate, December 8, 2001 found at http://www.fbi.gov/news/testimony/al-qaeda-international accessed on June 29, 2014.

Castel, Albert. *Civil War Kansas: Reaping the Whirlwind*, Lawrence, Kansas: University Press of Kansas 1997.

Cohen, William A., *The Art of the Strategist: 10 Essential Principles for Leading Your Company to Victory*, New York, New York: AMACOM, 2004.

Cole, Ronald, "Operation Just Cause: Panama," Washington, DC: Joint History Office, Office of the Chairman of the Joint Chiefs of Staff, 1995.

Cooke, John Esten, *Wearing of the Gray*, Whitefish, Montana: Kessinger Publishing, LLC, 2010.

Darby, William O. and Baumer, *William H., Darby's Rangers: We Led The Way*, California, Presidio Press, 1980.

DePuy, William, *Changing an Army*, Washington, DC: Center of Military History, 1988 found at http://www.history.army.mil/html/books/070/70-23/CMH_Pub_70-23.pdf.

Drea, Edward J., "Tradition and Circumstance: The Imperial Japanese Army's Tactical Response to Khalkin-Gol, 1939," in Colonel Charles R. Shrader, ed., *Proceedings of the 1982 International Military History Symposium: The Impact of Unsuccessful Military*

Campaigns on Military Institutions, 1860-1980, Washington, DC: US Army Center of Military History, 1984.

Drake, Samuel Adams, *The Border Wars of New England*, Charleston, South Carolina: Nabu Press, 2010.

Dupont, Jean-Claude, *Héritage d'Acadie*, Montreal, Canada: Éditions Leméac, 1977.

Eckhardt, G.S., preface to the "Study on Military Professionalism," Carlisle Barracks, Pennsylvania: US Army War College, June 30, 1970.

Encyclopedia of Oklahoma History and Culture, "Quantrill's Raiders," found at http://digital-library.okstate.edu/encyclopedia/entries/Q/Q002.htm, accessed on February 10, 2011.

Evans, Thomas W., "The All-Volunteer Army after Twenty Years: Recruiting in the Modern Era," Huntsville, Texas: Sam Houston State University, Summer 1993 found at http://www.shsu.edu/~his_ncp/VolArm.html accessed on May 5, 2010.

Faragher, John Mack, *A Great and Noble Scheme: The Tragic Story of the Expulsion of the French Acadians from their American Homeland*. New York, New York: W. W. Norton, 2005.

Ferguson, David, "Bush Says Slow Reaction on 9/11 was Deliberate," *The Raw Story,* June 29, 2011 found at http://www.rawstory.com/rs/2011/07/29/bush-says-slow-reaction-on-911-was-deliberate-decision accessed on June 9, 2012.

Field Manual (FM) 7-8, Infantry Rifle Platoon and Squad, Washington, DC: Department of the Army, April 22, 1992.

Flanagan, Edward M., *Battle for Panama, Inside Operation Just Cause*, Dulles, Virginia: Brassey's (US) Inc., 1993.

Foner, Eric, *Reconstruction: America's Unfinished Revolution, 1863–1877*, The New American Nation series, New York: New York: Harper & Row, 2008.

Fortnight, David, *Crisis in Command: Mismanagement in the Army*, New York, New York: Hill and Wang, 1978.

Franks, Tommy, *American Soldier*, New York, New York: Harper Collins, 2004.

Frink, Tim, *New Brunswick, A Short History*. Summerville, New Brunswick: Stonington Books, 1999.

Richard A. Gabriel, A., *Military Incompetence: Why the American Military Doesn't Win*, New York, New York: Hill and Wang, 1985.

Gabriel, Richard A. and Paul L. Savage, *Soldiers in Revolt: The American Military Today*, Garden City, New York: Doubleday, 1975.

Geier, Joel, "Vietnam: The Soldier's Revolt," *International Socialist Review* Issue 9, August-September 2000 found at http://www.isreview.org/issues/09/soldiers_revolt.shtml accessed on March 3, 2012.

Gresham, John D., "The Haditha Dam Seizure – The Target," *defensemedianetwork*, May 1, 2010 found at http://www.defensemedianetwork.com/stories/hold-until-relieved-the-haditha-dam-seizure/ accessed on June 9, 2012, and July 2, 2014.

Griffith, Robert K, Griffith, Robert K., Jr. and Wyndham Mountcastle, John, *U.S. Army's Transition to the All-volunteer Force, 1868-1974*, Darby, Pennsylvania: Diane Publishing, 1997.

Grenier, John, *The First Way of War: American War Making on the Frontier*, Cambridge, Massachusetts: Cambridge University Press, 2005.

Gray, David R., *Black and Gold Warriors: US Army Rangers During The Korean War* Michigan: University Microfilms International, 1992.

Heinl, Robert D., Jr., "The Collapse of the Armed Forces," *Armed Forces Journal* (June 7, 1971), p. 1-2 found at http://chass.montclair.edu/english/furr/Vietnam/heinl.html accessed on June 12, 2014.

Hopkins, James E. T. and Jones, John M., *Spearhead: A Complete History of Merrill's Marauder Rangers,* Baltimore, Maryland: Galahad Press, 1999.

Improving the Prospects for Future International Peace Operations, Washington, DC: Office of Technology Assessment, 1995.

Indian Narratives, Claremont, New Hampshire: Tracy and Brothers, 1854.

Jehl, Douglas, "Ranger Force Bore Brunt of Panama Toll," *Los Angeles Times*, January 7, 1990 found at http://articles.latimes.com/1990-01-07/news/mn-445_1_rio-hato/2 accessed on May 31, 2012.

Johnson, Frank, Diary of an Airborne Ranger: A LRRPs Year in the Combat Zone, New York, New York: Ballantine Books. 2001.

Jung, Patrick J, *The Black Hawk War of 1832,* Norman, Oklahoma: University of Oklahoma Press, 2008.

Keen, Ken, "The Ranger Regiment: Strategic Force for the 21st Century: Strategy Research Project, Carlisle Barracks, Pennsylvania: US Army War College, April 1, 1998.

Kearney, Francis H. III, "The impact of leaders on organizational culture: a 75th ranger regiment case study," Carlisle Barracks, Pennsylvania: *US Army War College,* 1997.

King, Michael J., *Rangers: Selected Combat Operations in World War II,* Fort Leavenworth, Kansas: Combat Studies Institute, US Army Command and General Staff College, June 1985.

Kinzer, Stephen, "Inside Iran's Fury, Scholars Trace the Nation's Antagonism to its History of Domination by Foreign Powers," *Smithsonian Magazine*, October 2008 found at http://www.smithsonianmag.com/people-places/iran-fury.html accessed on May 6, 2012.

Kirkham, R. R., Chamness M. A., Driver, C. J. and Barfuss, B. C., "Air Quality and Road Emission Results for Fort Stewart, Georgia," *Pacific Northwest National Laboratory* found at http://www.pnl.gov/main/publications/external/technical_reports/PNNL-14933.pdf (February 2005) assessed on June 18, 2014.

Kozary, Linda D., "n Defense Leaders Uphold Army's Black Beret Decision," *American Forces Press Service,* March 16, 2001 found at http://www.defense.gov/news/newsarticle. aspx?id=45776 accessed on June 29, 2104.

Ladd, James, *Commandos and Rangers of World War II*, New York, New York: St. Martin's Press, 1978.

Lewis, Jon E., ed., *The Mammoth Book of Special Forces: True Stories of the Fighting Elite Behind Enemy Lines*. Philadelphia, Pennsylvania: Running Press, 2004

Lock, John D., *ArmyRangers.Com: For Rangers by Rangers, Online since 1999.* "The Vietnam War," found at http://www.armyranger.com/index.php/history/modern-era/vietnam accessed on February 5.2012.

Lock, John D., *To Fight with Intrepidity: The Complete History of the US Army Rangers 1622 to Present,* New York, New York: Pocket Books, 1998.

Lock, John. D. *Rangers in Combat: A Legacy in Valor,* Tucson, Arizona: Wheatmark, 2007.

Lock, John D., *The Coveted Black and Gold: A Daily Journal Through the US Army Ranger School Experience,* Xlibris Corporation at www.Xlibris.com, 2001.

Mackey, Robert Russell, *The Uncivil War: Irregular Warfare in the Upper South, 1861-1865,* Norman, Oklahoma: University Oklahoma Press, 2004.

Maloney, Sean M., *Enduring the freedom: a rogue historian in Afghanistan*, Dulles, Virginia: Potomac Books, Inc., 2006.

Martin, Robert J. and Taylor, Thomas H., *Rangers Lead the Way*, Nashville, Tennessee: Turner Publishing, 1996.

Mason, Betsy, "The Science of the Future of War," *Wired Science,* November 28, 2008 found at http://www.wired.com/wiredscience/2008/11/sex-and-war-exc/ accessed on June 17, 2012.

McManners, Hugh, *Ultimate Special Forces: The Insider's Guide to the World's Most Deadly Commandos*. New York, New York: DK Publishing, 2006.

Mcraven, William, *Spec Ops: Case Studies in Special Operations Warfare: Theory and Practice*, New York, New York: Presidio Press, 2006.

Miller, Jeremy B., "Unconventional Warfare in the American Civil War," Fort Leavenworth, Kansas: Command and General Staff College, 2004.

Moore, Robin and Lennon, Michael, *The Wars of the Green Berets: Amazing Stories from Vietnam to the Present,* New York, New York: Skyhorse Publishing, 2007.

Forrest E. Morgan, Karl P. Mueller, Evan S. Medeiros, Kevin L. Pollpeter and Roger Cliff, *"Dangerous thresholds: managing escalations in the 21st century,"* Santa Monica, California: *Rand Corporation Project Air Force,* 2008.

Morris, Steven A., "Operation Iraqi Freedom, Objective Lynx, Haditha Dam," Fort Leavenworth, Kansas: *Combined Arms Research Library Digital Library*, June 10, 2009.

Mosher, Howard Frank, *North Country, A Personal Journey*. New York, New York: Houghton Mifflin, 1997.

Nadler, John, *A Perfect Hell: The True Story of the Black Devils, the True Forefathers of the Special Forces*, New York, New York: Ballantine Books, 2006.

Naylor, Sean, *Not a Good Day to Die; The Untold Story of Operation Anaconda*, New York, New York: Berkley/Penguin Publishing, 2005.

Neville, Leigh, Carpenter, Shawn, Wonacott, Jim, Jim Roots, and Carpenter, Robby, *Road to Baghdad: Iraq 2003,* Oxford, UK: Osprey Publishing, 2011.

Nielsen, Suzanne C., "An Army Transformed: The US Army's Post-Vietnam Recovery and the Dynamics of Change in Military Organizations," The Letort Papers, Carlisle Barracks, US Army War College, September 2010 found at http://www.strategicstud-iesinstitute.army.mil/pdffiles/PUB1020.pdf accessed on February 15, 2012.

North, Oliver, *American Heroes in Special Operations,* Jarrell, Texas: Fidelis 2010.

Ogburn, Charlton, *The Marauders*. New York, New York: Harper & Brother, 1956.

Paul, Ron, Forums: *Liberty Forest*, April 2003 found at http://www.ronpaulforums.com/showthread.php?454471-Haditha-Dam-Part-II accessed on July 3, 2014.

Ranger Handbook (SH 21-76) found at http://www.africom.mil/WO-NCO/DownloadCenter/%5C40Publications/Ranger%20Handbook.pdf accessed on December 31, 2011.

Rankin, Nicholas, Churchill's Wizards: The British Genius for Deception 1914–1945. Mechanicsburg, Pennsylvania: Stackpole Books, 2008.

Rashid, Ahmed, *Taliban: Militant Islam, Oil & Fundamentalism in Central Asia,* New Haven, Connecticut: Nota Bene Yale University Press, 2001.

Rhem, Kathleen T., "Bush: No distinction between attackers and those who harbor them," American Forces Press Service, Washington DC, September 11, 2001 found at http://www.defense.gov/news/newsarticle.aspx?id=44910 accessed on June 9, 2012.

Rinaldi Matthew, "The Olive-Drab Rebels: Military Organizing during the Vietnam Era," *Radical America*, Vol.8 No. 3, May-June 1974.

Ritter, Charles F. and Wakelin, Jon L. *Leaders of the American Civil War A Biographical and Historiographical Dictionary*, Portsmouth, New Hampshire: Greenwood Publishing Group, 1998.

Rosen, Stephen Peter, "The Future of War and the American Military," *Harvard Magazine*, May-June 2002 founds at http://harvardmagazine.com/2002/05/the-future-of-war-and-th.html accessed June 17, 2012.

Rosenbaum, David E., "Senate Approves Draft Bill, 55-30; President to Sign," *The New York Times*, (1971-09-22) found at http://select.nytimes.com/mem/archive/pdf?res=F50B10F D385C1A7493C0AB1782D85F458785F9 accessed on December 12, 2011.

Ross, John F., *War on the Run: The Epic Story of Robert Rogers and the Conquest of America's First Frontier*, New York, New York: Bantam Books, 2009.

Schmitt, Eric, "Threats and responses: military spending: Pentagon contradicts general on Iraq occupation force's size," *New York Times*, February 23, 2003 found at http://www.nytimes.com/2003/02/28/us/threats-responses-military-spending-pentagon-contra-dicts-general-iraq-occupation.html, accessed on June 30, 2014.

Scott, Michaud, "History of the Madawask Acadians," found at http://members.tripod.com/~Scott_Michaud/Madawaska-history.html accessed on March 5, 2008.

Sizer, Mona D., *The Glory Guys: The Story of the US Army Ranger*, Lanham, Maryland: Taylor Trade Publishing 2005.

Skow, John, "The Long Ordeal of the Hostages," *Time*, January 26, 1981 found at http://www.time.com/time/magazine/article/0,9171,954605-3,00.html accessed on May 6, 2012.

Sorley, Lewis, "The Art of Taking Charge," *Across the Board*, May 1992.

Spector, Ronald H., *US Marines in Grenada 1983*, Washington DC: *History and Museums Division*, 1987.

Stanton, Shelby, *Rangers at War: Combat Recon in Vietnam*, New York, New York: Presidio Press, 1992.

Stanton, Shelby L., *Vietnam Order of Battle*, Mechanicsburg, Pennsylvania: Stackpole Books, 2003.

Stewart, Richard W., Editor, *American Military History: The United States Army in a Global War, 1917-2003, Volume II*, Washington, DC: Center of Military History, "Rebuilding the Army: Vietnam to Desert Storm," 2005 found at http://www.history.army.mil/books/AMH-V2/AMH%20V2/chapter12.htm accessed on February 15, 2012.

Stone, Andrea, "Military's aid and comfort ease 9/11 survivors' burden," *USA Today*, August 20, 2002.

Swenson, Derek, *In the Eyes of the Beholder*, Xlibris found at http://books.google.com/books?id=8WJNAAAAQBAJ&pg=PT220&lpg=PT220&dq=helicopters+lifted+off+at+1532.&source=bl&ots=Hs76GynPuv&sig=Jztn9zBOE8QCJyVb5-w1rtBTyso&hl=en&sa=X&ei=RV6sU9iQH9CoyATL3oCYDA&ved=0CEYQ6AEwB A#v=onepage&q=helicopters%20lifted%20off%20at%201532.&f=false (2010).

Todish, Timothy J. and Gary S. Zaboly, The *Annotated and Illustrated Journals of Major Robert Roger*, Fleischmanns, New York: Purple Mountain Press, 2002.

Tsouras, Peter G. *The Greenhill Dictionary of Military Quotations*, Mechanicsburg, Pennsylvania: Greenhill Books, 2000.

Tsouras, Peter G, *Warrior's Words: A Quotation Book From Sesostris III to Schwarzkopf 1871 BC to AD 1991,* London: Arms and Armour Press, 2006.

US Special Operations Forces, Washington, DC: *US International Business Publications,* 2011.

Urwin, Gregory J. W. *The United States Cavalry: An Illustrated History, 1776-1944,* Norman, Oklahoma: University of Oklahoma Press, 1983.

Wellenson, Kim, "Drugs 2-22," *Wire Services News,* Washington DC, February 22, 1973.

Wert, Jeffry D., *Cavalryman of the Lost Cause: A Biography of J. E. B. Stuart,* New York, New York: Simon & Schuster, 2009.

Woods, Kent T., "Rangers Lead the Way: The Vision of General Creighton W. Abrams," US Army War College Strategy Research Project, Carlisle Barracks, Pennsylvania, July, 14, 2003.

Woodward, Bob, *Plan of Attack,* New York, New York: Simon & Schuster, 2004.

Zaboly, Gary S., *The Annotated and Illustrated Journals of Major Robert Rogers,* Utica, New York: Purple Mountain Pr Ltd, 2002.

Zirin, David, *People's History of Sports in the United States: 250 Years of Politics, Protest, People, and Play,"* New York, New York: The New Press, 2009.

Zinn, Howard, A People's History of the United States: 1492-Present, New York, New York: Harper Perennial, 2005, Extract found at http://libcom.org/history/1983-the-us-invasion-of-grenada, accessed on May 10, 2012.

DOCUMENTS AND PAPERS

American Archive Documents of the American Revolution, 1774-1776, produced by Northern Illinois University Library p. V2:1847 found at http://lincoln.lib.niu.edu/cgi-bin/amarch/getdoc.pl?/var/lib/philogic/databases/amarch/.5774

Baker, John , "SUBJECT: Personnel Actions," Memorandum to the Commander, XVIII Airborne Corps and Fort Bragg, April 9, 1974, Military History Institute, Carlisle Barracks, Creighton W. Abrams Papers.

Bill Gertz, "Aspin's decision on tanks was political," in *The Washington Times* (October 3) found at http://www.sportsjournalists.com/forum/index.php?topic=32177.135;wap2 1995) accessed on June 2, 2012.

Commander, Military Personnel Center, Subject: Application for Assignment to Ranger Duty. AUTODIN Message to AIG 7401, Alexandria, Virginia, March 5, 1973.

General Creighton W. Abrams notes from Staff Meeting Military History Institute, Carlisle Barracks, Pennsylvania, February 26, 1973.

Ivers, Larry E. "A Brief History of the American Rangers: 1634-1990," *United States Army Ranger Association, Inc.* date of publication unknown.

Lineage and Honors, 75th Ranger Regiment, Washington DC: The Department of the Army, March 26, 2013 found at http://www.history.army.mil/html/forcestruc/lineages/branches/inf/0075ra.htm accessed on July 7, 2014.

Memorandum for Record: Army Training Test Concept dated January 7, 1974 found at http://usacac.army.mil/cac2/CSI/docs/Gorman/03_DCST_1973_77/04_74_ArmyTngTests_7Jan.pdf, accessed on March 23, 2012

Secretary of the Army Howard H. Callaway, letter to General Creighton W. Abrams, Washington, DC, August 22, 1974, Military History Institute, Carlisle Barracks, Pennsylvania, Creighton W. Abrams Papers.

United Nations General Assembly resolution 38/7, page 19," United Nations, November 2, 1983, found at http://www.un.org/depts/dhl/resguide/r38.htm, accessed on May 11, 2012.

United Nations Yearbook, Volume 37, 1983, Department of Public Information, United Nations, New York.

US Army War College, "Study on Military Professionalism," Carlisle Barracks, Pennsylvania, 1970.

Webster, Daniel (1814-12-09) "On Conscription," reprinted in *Left and Right: A Journal of Libertarian Thought* (Autumn 1965).

William and Mary Quarterly (July 1923)

Wilson, Jeffrey S, "Transformational Leadership: William Depuy's Vision for the Army," *Military Review* (September 1, 2011), found at http://readperiodicals.com/201109/248091260.html.

INTERVIEWS

While researching and preparing for this book, I conducted numerous, and often repetitive, interviews with many great Americans. I also relied on a number of individuals that conducted the interviews for the work like Keith Nightingale, Ron Rokosz and Tom Currie. I do not reference all these interviews as they are cited in the narrative of this work and it would be cumbersome and a distraction to the reader. Many of the stories are provided by a number of Rangers present and past and the acknowledgment of their input is imbedded in the work.

In each entry I mention whose thoughts I cite. I do provide references to specific documents and augment as necessary to any other sources used while researching those topics.

The following individuals provided support in some fashion (interviews, reference, guidance, etc.) during the preparation of this work:

Staff Sergeant (retired) Bill Acebes B Company
Colonel (retired) Jerry Barnhill, the original C Company Commander
CW4 (retired) Chris Brewer, B Company
Command Sergeant Major (retired) Jimmy Broyles, C Company
Staff Sergeant (retired) Don Bruce, C Company
Staff Sergeant (retired) Mike Cheney, A Company
Captain (retired) Don K. Clark, original A Company Commander
Major (retired) Todd Currie, Ranger, US Army and one of the leaders of this project
Colonel (retired) Gerry Cummins who served with the US Army Rangers from 1978-1981 on March 4, 2012.
Lieutenant General (retired) Karl Eikenberry, platoon leader B Company and former US Ambassador to Afghanistan.
Sergeant Jeff Everett, A Company
Sergeant Ron Fallon, B Company
Master (retired) Sergeant Don Feeney, C Company
Private Joe Hill, A Company
Command Sergeant Major (retired) George Horvath, B Company
Colonel Dave LeMauk, B Company

Brigadier General William J. Leszczynski, Jr., the 9th Colonel of the 75th Ranger Regiment about the heights of the cliffs that he attained while serving on the American Battles Monument Commission. Interview on December 10, 2011

Major General (retired) K.C. Leuer, the first Ranger Battalion Commander

Lieutenant General (retired) Lawson G. Magruder, telephone interview by Kent T. Woods, March 5, 2003.

Command Sergeant Major (retired) Frank Magaña, B Company

First Sergeant (retired) Mike Martin, the original 1SG for B Company

Command Sergeant Major (retired) Joe Mattison, B Company

Lieutenant Colonel (retired) Jim Moeller, B Company

Colonel (retired) James Montano, Battalion S5

Colonel (retired) Scot Newport interviewed on December 1, 2011l: Infusion of intelligence and operations habitually found as a norm in special operations forces (SOF) of which the Rangers are a large part became prevalent throughout the big Army both through infusion of Rangers and other special operations forces into the Big Army and the introduction of the conventional force brought on by the wars in the Middle East where the lines of SOF and other regular forces fight in the same realm.

Colonel (retired) Keith Nightingale, the original HHC Commander and another leader in this project.

Private First Class Jerry Olsen, B Company

Lieutenant Colonel (retired) Jack Rogers, B Company

First Sergeant (retired) Steve Rondeau

General (retired) William R. Richardson, the Assistant Commander of Fort Benning at the time of the forming of the Rangers.

Brigadier General (retired) Ron Rokosz, the original B Company Commander and project leader of the interviews and driver behind much of what was accomplished in defining the purpose of this work.

Lieutenant General (retired) Gary Speer, original A Company Executive Officer

General (retired) Volney Warner, The Deputy Chief of Staff for Operations, US Army at the time of the forming of the first battalion of the 75th Rangers.

Sergeant First Class (retired) Mike Wagner, C Company.

WEBSITES

These are sites that are accessed periodically to gain in depth information on various topics. There are other websites listed throughout this bibliography that are specific to an article or summary found in each:

Arlington National Cemetery Website, "Creighton Williams Abrams, Jr.", available from http://www.arlingtoncemetery.com/abrams.htm.

http://darbysrangers.tripod.com/id4.htm

http://en.inforapid.org/index.php?search=Operation%20Rhino

http://en.wikipedia.org/wiki/Acadians

http://en.wikipedia.org/wiki/Lockheed_AC-130

http://en.wikipedia.org/wiki/Lockheed_P-3_Orion

http://mymilitaryhistorypages.bravehost.com/RangerRiflemenSketchbook.htm

www.psywarriors.com/ranger.html.

http://www.rangerfamily.org/Commanders/Wm%20O%20Darby.htm

http://www.americanmilitaryhistorymsw.com/blog/620296-us-army-rangers-at-dieppe/

http://www.army.mil/features/beret/beret.htm

http://www.armyranger.com/index.php/history/early-years/american-revolution

www.biographyonline.net./sport/muhammad_ali.html

http://www.corydonbattlepark.com/battle.html

http://www.fideles.net/fidelesrangers.cfm

http://www.funker530.com/ac-130-gunship/

http://www.historicmosbysrangers.org/

http://www.history.army.mil/documents/RevWar/revra.htm

https://www.hrc.army.mil/site/Active/TAGD/awards/Foreign_Badges_Appendix/
    Appendix_D.pdf

http://www.irelandseye.com/aarticles/history/events/worldwar/rangers2.shtm

http://www.johndlock.com/#!tfwi-extract---tf-ranger-somalia/crl8

http://www.msnbc.msn.com/id/4677978/ns/world_news-hunt_for_al_qaida/t/al-qaida-
    timeline-plots-attacks/

http://www.patriotshistoryusa.com/teaching-materials/bonus-materials/american-heroes-
    francis-marion/

http://www.rand.org/pubs/research_briefs/RB9195/index1.html

http://www.ranger.org/Default.aspx?pageId=578463

http://www.rangerfamily.org/History/History/Battalion%20Pages/fifth.htm

http://www.rangerfamily.org/History/History/Battalion%20Pages/sixth.htm

http://www.reference.com/browse/M163_VADS.

http://www.reuters.com/article/2011/01/21/us-iran-nuclear-sanctions-idUSTRE70K03O
    20110121, "Factbox: U.S. EU, and U.N. sanctions against Iran," Reuters Web Page,
    January 20, 2011.

http://www.shsu.edu/~his_ncp/VolArm.html

http://www.sfa-72.com/html/moh_gary_gordon-_hon_life_member_sfa-72.html

http://www.smithsonianmag.com/history-archaeology/biography/fox.html#ixzz1jpKyjzwW

http://www.socnet.com/showthread.php?t=2736

http://www.stuart-mosby.com/laura-ratcliffe.

http://www.sss.gov/lotter1.htm

http://www.suasponte.com

http://www.suasponte.com/m_beret.htm

http://www.suasponte.com/vietnam.htm

http://www.smithsonianmag.com/history-archaeology/biography/fox.html#ixzz1jjkR8oOd

http://www.smithsonianmag.com/history-archaeology/biography/fox.html#ixzz1jpKyjzwW

http://www.whitehouse.gov/news/releases/2001/10/2001 1007-8.html.

http://www.wwiirangers.com/History/History/Battalion%20Pages/sixth.htm, Contribution
    on Ranger World War II History website by Leo V. Strausbaugh (Colonel, US Army
    retired).

http://www.youtube.com/watch?v=fcF2ctMdMsE

Index

INDEX OF PEOPLE

INDEX OF PLACES

INDEX OF MILITARY UNITS

INDEX OF MISCELLANEOUS/OTHER TERMS